KT-458-843

PLAYING GOD?

Playing God?

Genetic Determinism
and Human Freedom

Ted Peters

ROUTLEDGE
New York and London

Hertfordshire Libraries, Arts and Information

H31 684 692 7	
BC	5\97
	£40.00

Published in 1997 by

Routledge
29 West 35th Street
New York, NY 10001

Published in Great Britain in 1997 by

Routledge
11 New Fetter Lane
London EC4P 4EE

Copyright © 1997 by Routledge

Printed in the United States of America
Design: Jack Donner

All rights reserved. No part of this book may be reprinted or reproduced or utilized in any form or by any electronic, mechanical, or other means, now known or hereafter invented, including photo-copying and recording, or in any information storage or retrieval system without permission in writing from the publishers.

Library of Congress Cataloging-in-Publication Data

Peters, Ted, 1941 –
    Playing God? : genetic determinism and human freedom / by Ted Peters.
        p.   cm.
    Includes bibliographical references and index.
    ISBN 0–415–91521–X (alk. paper). — ISBN 0–415–91522–8 (pbk. : alk. paper)
        1. Human genetics—Moral and ethical aspects.   2. Human genetics—Religious aspects—Christianity.   3. Medical ethics.   4. Christian ethics.   5. Free will and determinism.   I. Title.
    QH438.7.P48   1996
    174'.9574—dc20   96–32892
                                                                                         CIP

Contents

Acknowledgments

Opportunity to pursue the background research for this book began with my work as principal investigator on a grant from the United States National Institutes of Health dealing with "Theological and Ethical Questions Raised by the Human Genome Project." I am grateful to the Ethical, Legal, and Social Issues divison of the NIH for their support from 1991 to 1994.

My friends and colleagues at the Center for Theology and the Natural Sciences, which is an affiliate of the Graduate Theological Union in Berkeley, California, have provided me with a congenial and encouraging atmosphere. Whatever value my contributions to the growing field of Theology and Science might turn out to be is attributable to the vision and creativity and excitement of this research center.

Particular individuals have helped me in preparation of this manuscript by giving me critical feedback or encouragement or both. I would like to thank Susan Ashbourne, Patricia Codron, Greg Cootsona, Elizabeth Cushing, Lindon Eaves, Greg Hile, Carol Jacobson, David Peters, Erika Peters, Mark Richardson, Robert John Russell, and Lisa Stenmark, and especially Stan Lanier for indexing.

Ted Peters
Berkeley, California
June 12, 1996

Foreword

The two words "playing God" are powerful. In the past they were directed primarily at the activities of medical personnel involved in dramatic interventions or end-of-life decisions. Today these two words are increasingly finding their way into discussions of the consequences of genetic research. Interestingly, the person uttering these words will often not otherwise acknowledge the existence of a God who is concerned with human beings—but the interjection of this phrase predictably results in elevated pulses and blood pressures, as the defenders of genetic research seek to justify its ultimate power to do good, while the critics seek to demonstrate how this approach crosses lines that the human species should not dare to traverse.

The mandate to alleviate human suffering is one of the most compelling of all expectations of humanity. Jesus Christ Himself spent a remarkable fraction of his brief time on earth healing the sick. When genetics is seen to fall into that larger mandate, it is hard to argue with its potential goodness. In fact, given that potential, it can be argued that the most unethical approach of all would be to insist that genetic research be stopped; because if it were, those individuals, present and future, who suffer from the ravages of genetic disease would be doomed to hopelessness. Given that virtually all diseases have a genetic component, and that we all carry predispositions to certain illnesses, the hope of a healthier future through genetic research is not restricted to a rare individual here and there; it applies to all of us.

Yet when genetics moves out of the realm of disease and into the study of human traits, especially when an intent to alter traits is implied or openly stated, the discomfort level appropriately rises, and questions about "playing God" are often raised. Perhaps if we could be confident that humans would play God as God does—with infinite love and compassion—the concern would be lessened. The fear expressed in the phrase "play God," however, is that humans might play God in their own selfish and imperfect ways.

There is a widespread assumption in much of modern Western culture that science and faith have settled into a position of unresolvable opposition on many issues, and genetics is often cited as a cardinal example. In that context, it may be surprising to come to learn that the Director of the National Center for Human Genome Research is also a serious Christian. I came to my own conclusions about faith at age twenty-seven, after several years of agnosticism bordering on atheism were brought to a sudden end when I encountered the works of C. S. Lewis. My own faith is not based on childhood exposure or emotional experience, but rather on the kind of logical argument for the reasonableness of Christianity which Lewis presents so well. I find no discordance between being a scientist who insists on absolute rigor in studying the natural world and being a person of faith who believes in a personal God. In that context, I find those rare dramatic moments of scientific discovery in my own experience to be moments of worship also, where a revelation about some new intricacy of God's creation is appreciated for the first time.

As genetic predispositions to everything from cancer or diabetes to novelty-seeking behavior or homosexuality are being reported almost daily in the scientific literature (and regrettably often overstated in the popular press), a new and dangerous brand of genetic determinism is subtly invading our culture. Carried to its extreme, this "Genes R Us" mentality would deny the value of social interventions to maximize individual potential, destabilize many of our institutions (perhaps especially the criminal justice system), and even deny the existence of free will. Surely a world in which every aspect of human behavior is hard-wired into our genes cannot comfortably exist with the concept of personal responsibility and free will to try (albeit not successfully for very long) to follow the moral law of right and wrong which people of faith believe has been written into our hearts by a loving and holy God.

In this remarkable book, Ted Peters explores the fallacies of the "gene myth" and presents a resounding array of arguments against this kind of all-encompassing genetic determinism. On the scientific side, he correctly points out that genetic influences on behavior are in most instances relatively modest. Does anyone deny that identical twins are still able to practice individual free will? After dispatching some of the sweepingly deterministic conclusions of the "science" of evolutionary psychology with a particularly effective set of rebuttals, Peters arrives at the conclusion that the nature of humanness is an interaction of three things—not one, not two, but three: genetics, environment, and free will. Neglecting any one of these three leads down a path of fuzzy thinking and dangerous consequences.

It is my hope that scientists, philosophers, and theologians will pay close attention to these arguments. There is no doubt that significant challenges lie ahead in the proper application of the powerful revelations of human genetics. Right now, for instance, an urgent need exists to provide protections to prevent the use of genetic information to deny health insurance or employment, a situation that should be a moral outrage. The Church, with its powerful tradition of effective advocacy on moral issues, can contribute much to the navigation of these difficult waters. Yet often the Church's recent forays into genetics, whether in murky statements decrying patenting of genes or in articles venting unnecessary moral outrage at unrealistic mad-scientist scenarios, have not inspired confidence or increased credibility. Good theology needs accurate science. Proverbs 19:2 exhorts, "It is not good to have zeal without knowledge." This volume is a great place to start. *Playing God? Genetic Determinism and Human Freedom* is an effective and well-reasoned presentation of many of the most crucial scientific, ethical, and theological issues that face all of us as we move into a new, promising, but potentially unsettling era of molecular medicine.

—*Francis S. Collins, M.D., Ph.D.*
Director, National Center
for Human Genome Research

Preface

The hurricane speed of genetic discoveries and the driving winds of ethical turbulence seem to be blowing our once secure notions of human freedom off their foundations. Like a weather forecast, this book tracks the storm as it moves from the laboratories of the molecular biologists through the region of the gene myth until it slams against the mainland of previously established beliefs.

As science filters into everyday society, it is being interpreted according to the *gene myth*, a cultural thought form that says, "It's all in the genes!" This growing myth of genetic determinism blows first in one direction: if we are programmed totally by our DNA, then what we think is human freedom is in fact a delusion. Then the myth blows the opposite way: if we can apply our best engineering technology to DNA, then we can gain control over nature and guide our own evolutionary future. The genes determine the future; we want to determine the genes. But should we? Are there no trespassing signs saying, "Don't play God with our DNA?"

This book examines the concept of "playing God" that arises within the genetic controversy and draws upon distinctively theological resources for understanding human creativity in light of divine creativity and the promise of redemption. It confronts ethical challenges to moral responsibility when genes predispose us toward socially unapproved behavior such as crime or alcoholism. It assesses the degree to which ethics should rely upon science when incorporating the discovery of the so-called "gay gene." It examines the public outcry against gene patenting. It debates the issue of germline intervention, wherein today's engineering will affect the genome of future generations.

The gene myth is misleading and unhealthy. In this book I will argue that a theological understanding of the human being as future-oriented and cocreative will lead to a healthier ethic for guiding genetic research and its technical applications. Genetic knowledge will not eliminate our confidence in human freedom, nor in the moral responsibility entailed by such freedom. Genuine freedom, especially Christian freedom, expresses itself as beneficence in the use of science to relieve human suffering and to make this a better world in which to live.

Our analysis begins at the point where science intersects culture. In a sense, science is an expression of culture. In the post-Enlightenment West, science is the way culture produces knowledge. The kind of knowledge it produces is born out of a cultural vision of human nature and human aspiration. Yet, science, like a teenager leaving home for college, learns things not previously anticipated by its parent. It gains critical distance and returns home with an altered view of reality. Science criticizes the culture it has inherited; it helps procreate the next generation of culture—that is, science contributes to the growth of tradition.

Such growth includes growing pains. Like the teenager who feels misunderstood by old-fashioned parents, the lab scientist frequently feels misunderstood by the media that report on his or her work. Not that the media are old-fashioned; rather, the media along with our educational institutions, commercial and political enterprises, and the arts have an investment in how new scientific knowledge gets interpreted. This investment need not always take the form of power-seeking or economic interest. This investment is of a conceptual nature as well. Our culture has an investment in certain identifiable conceptual sets or thought forms or subsystems of belief. These conceptual sets incorporate our fears and our hopes and frame our picture of reality. Through this frame we view new experiences, even new knowledge gained from scientific research. These conceptual sets are by no means fixed or static; they have an effective history that changes over time. Yet, they influence heavily the way in which scientific discovery becomes the common property of modern Western culture.

The millennia-old squabble over determinism and freedom is included in the conceptual set through which we interpret advances in molecular biology and other forms of genetic research. Although historically speaking this was originally a battle between theologians arguing about the efficacy of divine grace to determine salvation with or against human willingness, in its secularized Enlightenment form it is today fundamental to every dimension of the modern mind. The seventeenth-century Western mind split objects from subjects, and conveniently divided reality into what is determined and what must remain free. Nature is determined, and science studies nature as an object. People are free, because they are subjects. Everything having to do with human subjectivity falls on the free side of the ledger. So when objectivist science appears to be encroaching on the territory claimed by subjectivity, we revolt. Or, more accurately, we first feel alienated. Then we revolt, demanding liberation.

This tension between objectivity and subjectivity, between determinism and freedom, underlies every attempt on the part of our culture to interpret the Human Genome Project. Known as HGP, this worldwide research program enlists scientists from numerous nations in a common search to sequence the three billion nucleotides on the DNA and map the 100,000 genes. En route, molecular biologists are locating defective genes—the genetic predispositions to debilitating human diseases. Can we expect them to locate as well genes that will be determinative of human behavior? Even genes that influence us in the direction of antisocial behavior such as crime? If so, does this mean that what we experience as free will is a delusion? Is science taking away our freedom? Do we need a cultural backlash in order to liberate human subjectivity once again from the oppression of objectivist science?

This cultural struggle is being fought at the level of the gene myth, a conceptual set through which new scientific discoveries are being framed and viewed. It reflects the struggle in its nearly self-contradictory double assumption regarding genetic determinism: on the one hand, the genes govern us like a puppeteer, while on the other hand, once we gain control of the strings, we can become our own puppeteers.

All of this is fascinating to a theologian such as myself. My field is systematic theology, and within this field I specialize in theology of culture. I zero in on the point where natural science intersects with cultural values. Here, as in some of my former works, my method is akin to Langdon Gilkey's hermeneutic of secular experience—that is, an analysis of concepts in secular discussion to see what religious dimensions might be present.[1] Gilkey, now emeritus professor of systematic theology at the University of Chicago, is one of the pioneers in the contemporary theology-science dialogue. My approach will be to examine the gene myth, seeking to uncover hidden or implicit philosophical and theological concerns. Specifically, I will try to uncover concerns regarding the nature of human nature, with a special eye toward the dynamics of human freedom and responsibility.

We might call this a form of response theology. It is a form of intellectual discourse that responds theologically to issues prompted by public debate over scientific matters. This approach to theological matters is one I share with Robert John Russell and other colleagues at the Center for Theology and the Natural Sciences (CTNS) located at the Graduate Theological Union in Berkeley, California. This research and teaching center has borrowed methods of inquiry from the natural sciences and put them to use in theological research programs. Key here is the notion of *hypothesis*. Like scientists, theologians working at CTNS spell out the scholarly task in terms of hypothetical proposals, seeking to evaluate ideas and evidence. The mutual interaction of the natural sciences with academic theology promises new understandings for both fields, so we need a method that is open to growth in understanding. My response is one that aims to explain the relationship between theology, genetic science, and culture.

This may be a bit confusing to a reader who operates with an outdated conceptual set regarding what theology ought to be. If one assumes that theology is an anachronistic throwback to premodern dogmatism, then response theology may seem new or odd. My nonauthoritarian approach to theology would certainly be confusing to some contemporary science writers such as Carl Sagan, whose understanding of theology is limited to the ancient prescientific era. When railing against religion in his recent book *The Demon-Haunted World*, he reports that "the organized religions do not inspire me with confidence. . . . The fact that so little of the findings of modern science is prefigured in Scripture to my mind casts further doubt on its divine inspiration."[2] That any literature of the ancient world, whether religious or scientific, would lack information on what we have learned scientifically in only the last few centuries should surprise no one. Yet Sagan assumes that he can compare modern scientific methods and knowledge with a two-thousand-year-old Bible and easily dismiss the latter for its ignorance about the natural world. Because he does not engage *contemporary* intellectuals who have theological acumen, he feels he can ask rhetorically, "What sermons even-handedly examine the God-hypothesis?"[3]

The mutual interaction between science and theology can yield new and broader understanding only if scientists and theologians are willing to engage in serious dialogue and reflection. This requires patience and intellectual empathy as representatives of each field try to apprehend better what is going on within the other. This is the agenda that the scientists and religion scholars working with the Center for Theology and the Natural Sciences have set.

To the scientists, I ask for openmindedness and a willingness to engage in honest dialogue with theologians and other interpreters of culture to gain broader understanding about the impact scientific research is having on life in contemporary society. To the theologians, I ask again for openmindedness and willingness to engage in reflection regarding the implications for understanding reality prompted by natural science. The words of theologian Anne Clifford provide direction here. "Theologians who engage in dialogue with scientists and the theories of science are performing an important service for the community of believers. It is desirable for theologians to explicitly attend to science in doing theology because the influence of science is a major factor in determining the meaning and validity of religious discourse."[4]

As we proceed, the reader will note the limited scope of this book. There are some relevant areas of study that we could include but do not. One such item is the current discussion regarding the relationship of brain to mind, wherein among other things the question of biological determinism reappears in a way that parallels genetic determinism. Another important item is the history of eugenics in Germany during the Nazi period, a history that is not simply past, because it continues to animate lively political discussion over genetic research in central Europe. As important and as relevant as these topics are, I have elected to focus on the relationship of the gene myth to the Human Genome Project and associated forms of genetic research.

In sum, *Playing God?* examines the concept of genetic determinism, finding belief in human determinism unsupportable by science yet prevalent in our culture in the form of the "gene myth." Despite DNA determinism, we as persons are still free. We are also morally responsible. That responsibility includes building a better future through genetic science, a form of human creativity expressive of the image of God imparted by the divine to the human race.

< ONE >

Playing God
with DNA

I once for Freedom madly did aspire,
And stormed His bars in many a burst of rage:
But see, my Keeper with his brands of fire
Has cowed me quiet . . . and bade me love my cage!

—Arthur Stringer (Canadian), 1874–1948

Is DNA sacred? Is the human genome hallowed by nature? Is the genetic code at work in each of our cells a product of divine creation? Are our genes put there by God? If so, do we have the permission of nature or the permission of God to engineer our genetic code? If we broke into our own DNA with wrenches and screwdrivers in order to redesign ourselves, would we be violating something sacred? Would we the creatures become our own creators? Would we be playing God? Is it a sin to play God when we in fact are not God?

These are profound ethical questions. These ethical questions are prompted by a set of ontological questions regarding the nature of human being, questions asked of us by the new genetics.

The most threatening question is this: Are we about to lose our freedom? Or, to put it a bit more precisely: Will new discoveries in genetic science so completely explain human behavior that the freedom we previously thought we had will turn out to be a delusion? Do genes determine everything about us? Can we still think of ourselves as unique individuals? Can I rightly say, "It's all in my genes"? And, if we say that it's all in our genes, then should we surrender to fatalism, or should we seek some strength of will to use genetic knowledge to make the human lot a better one?

These ethical and ontological questions are also theological questions. Just the frequently used phrase "playing God" is sufficient reason to draw theologians into the contemporary discussion over genetics. This phrase, "playing God," is not found in standard theological dictionaries. It belongs to common parlance and is heard frequently around hospitals. It is heard still more frequently these days in discussions regarding genetic determinism. In this chapter we will tease out its meanings, both in common parlance and in theology.

We will begin by identifying the context, namely, the explosion of excitement over the Human Genome Project combined with the emerging discussion over genetic determinism. The public discussion of genetic determinism is not necessarily the same as the discussion among scientists, especially molecular biologists. The more popular level of discourse takes the form of what we will be calling the *gene myth*. We will identify two almost contradictory planks in the gene myth platform: puppet determinism and promethean determinism. We will also look at the gene myth's campaign slogan: Thou shalt not play God.

The gene myth both threatens and enhances human freedom. Here we will look at various understandings of freedom—political liberty, free will, moral freedom, and future freedom—and ask how each might be affected by the notion of genetic determinism. Within this framework we will look theologically at the concept of Christian freedom—a form of liberation that is given us by God's grace—along with its accompanying ethic of neighbor love described in terms of beneficence. Guided by the principle of neighbor love, we will ask the question of playing God in light of our responsibility to employ genetic science and its resulting technology in terms of loving our neighbor—that is, by reducing human suffering and enhancing the flourishing of the human race.

It will be my position that the phrase "playing God?" has very little cognitive value when looked at from the perspective of a theologian. Its primary role is that of a warning, such as the word "stop." In common parlance it has come to mean just that: stop. Within the gene myth it means we should stop trying to engineer DNA. Theologically, however, there is at best only minimum warrant for using this phrase in such a conservative and categorical way. Caution is always good advice, to be sure. Yet the task before us is to be good stewards of the advance of genetic science and technology so that it contributes to human welfare without creating new injustices.

What Is the Human Genome Project?

The worldwide excitement in genetics today is due to the leap-frogging advances in molecular biology. Workers at the laboratory bench and the computer terminal are producing knowledge about the biochemistry of human life at an unprecedented rate. This is due in large part to the dramatic impact of the Human Genome Project (HGP).

Begun in 1988 in the United States and soon followed by many other countries, the fifteen-year, three-billion-dollar Human Genome Project studies DNA with three goals in mind: sequencing, mapping, and diagnosing.[1] The first goal is to learn the sequence of the three billion base pairs or nucleotides that comprise the DNA chains in our forty-six chromosomes. The second is to locate the genes on a DNA map—that is, to locate the estimated 100,000 smaller sequences on the DNA chains that code for proteins and determine what kinds of bodies we have. The third goal is the one that draws public support in the form of government funding for research: the identification of those genes that predispose us to disease. At the beginning of HGP it was estimated that 5,000 or more human diseases are genetically based.[2] Finding those genes and developing therapies to counteract their effects holds immense promise for improving human health and well-being. A hoped-for result of HGP is the transformation of medicine. "Medicine will move from a reactive mode (curing patients already sick) to a preventive mode (keeping people well)," writes genetic researcher Leroy Hood. "Preventive medicine should enable most individuals to live a normal, healthy, and intellectually alert life without disease."[3]

Perhaps some words about DNA are in order here. DNA, short for deoxyribonucleic acid, is a long molecule stretched out in a chain of nucleotides. The chain links can be likened to letters in a sentence, and DNA to a text or code that tells our bodies what to do. The alphabet consists of four letters—A,T,C, and G—each standing for a nucleotide or base: Adenine, Thymine, Cytosine, and Guanine. They come in predictable pairs facing each other: A plus T or C plus G, no other combinations. Starting with this known alphabet, the task of the Human Genome Project is to learn the sequence of the letters and to read the text. The size of the text is enormous. The card catalogue for the DNA library requires enormous computer capacity.

Some sequence segments along the larger chain "speak" by producing proteins, and we refer to these segments as the genes. Genes account for only a fraction of the DNA. The long segments between genes seem silent, and the suspicion is that these sequences may have no function. So, the apparently nonfunctioning parts have been affectionately named "junk DNA."

Genes "express themselves," and this expression results in distinctive bodily traits. All the genes in a particular individual constitute his or her genotype, and all the physical traits constitute his or her phenotype. The phenotype is the result of gene products brought to expression in a given environment. Gene products include proteins that transport chemicals within the cell or throughout the body as well as proteins that play structural roles such as muscle building. Enzymes constitute the largest category of gene products. Enzymes convert food to energy and to structural materials. The design for the structure and function of every molecular part of our body is fixed by the genes.

Although DNA is a rather stable molecule, on rare occasions when it replicates itself a spontaneous change occurs. These changes, known as mutations, alter the alphabetical code in the genes. These mutant forms of genes are called allelomorphs or alleles. Such mutant alleles may be benign; yet sometimes the altered genetic code results in a defective protein or even the complete cessation of protein synthesis. This has a physical impact upon our bodies, changing our traits. For example, the healthy people among us have a gene that specifies the normal protein structure for hemoglobin, the red-blood-cell pigment. People suffering from some types of chronic anemia carry an allele form of the gene, which causes a defective hemoglobin protein that is unable to carry the normal amount of oxygen to the body cells.

Gene hunters are hot on the trail of alleles linked to human diseases. Some have already apprehended their prey. Genetic predispositions for many diseases have already been identified. The gene for Huntington's disease has been found on the top of chromosome 4. A cystic fibrosis allele has been found on chromosome 7, colon cancer on chromosome 2, and diabetes on chromosome 11. The first discovery of a mutant gene for inherited breast cancer, named BRCA1, was located on chromosome 17 in 1994, and BRCA2 on chromosome 13 in 1995. Alzheimer's disease is likely due to a combination of defective genes on more than one chromosome. Lou Gehrig's disease, childhood leukemia, fragile X mental retardation, and Duchenne's muscular dystrophy have locatable genetic origins.

In some cases, such as cystic fibrosis and others, the mutation consists of tandem repeats of certain nucleotide triplets. It's as if someone held down three keys on a word processor, causing a repeatable sequence to print again and again. The number of repeats varies, and in the case of cystic fibrosis, the more repeats, the more severe the disease.

The medical objective is to use this knowledge of defective genes to develop diagnostic procedures and appropriate therapies. One form of gene therapy is supplementing or replacing faulty genes with good ones. Differentiated cells from the target area of the body can be cultured in vitro, and a new properly functioning gene transduced into them. To date it is impossible to enter into a cell and snip out the faulty gene. So, such therapy typically consists of adding an additional gene that works. Once the cells have received their new gene, they are reintroduced into the body.

The somatic gene therapy described here has some limitations.[4] For example, already differentiated cells do not reproduce themselves. When they wear out, they die. They are replaced by cells with the mutant gene. Therefore, the effect of somatic therapy is transient, short-lived.

This has led some researchers to raise the prospect of germline therapy. The objective here would be to alter the defective genetic code of the germ cells—that is, the gametes, the father's sperm and mother's ova—prior to fertilization. This would insure that future children would be conceived with the functioning gene. The limitation here is obvious: germline intervention would benefit future generations, but it would not offer help for those suffering today. In addition, many ethicists raise doubts about using genetic engineering to influence the genotypes of the unborn who have no voice in such decision making. We will take up the knotty problem of germline intervention later. Right now we will follow the path from the scientist's lab bench to the cultural interpretation of what the scientist has learned. Namely, to the gene myth.

What Is the Gene Myth?

The cover of the January 17, 1994 issue of *Time* pictured a naked man with outstretched arms. The artist had positioned a giant double helix within his chest. The cover's large print read: "Genetics: The Future is Now." The medium-sized print read: "New breakthroughs can cure diseases and save lives, but how much should nature be engineered?"

To engineer something implies that what is to be engineered is mechanical, predictable. Are genes mechanical and predictable? Can we actually get control over genes as we can control the burning of fossil fuel in automobile engines? Can we govern the direction genes take the way we try to govern the course of flooding rivers when asking the Army Corps of Engineers to build levies? On the one hand, no. On the other hand, yes. These two apparently contradictory answers are components in the *gene myth*.

What we are talking about goes by a number of different names. John Maddox, writing for *Nature*, calls it the "Strong Genetic Principle."[5] Joseph Alper and Jonathan Beckwith refer to it as "genetic fatalism."[6] Dorothy Nelkin and Susan Lindee have named it "genetic essentialism."[7] I have elected to use the term *gene myth* as employed in the work of Ruth Hubbard and Elijah Wald, *Exploding the Gene Myth*.[8] No matter the term, we are talking about a thought form or conceptual set, a cultural frame through which we interpret the accelerating growth in scientific knowledge about DNA.

There is something special—something almost sacred—about the genes. In the early days of the Human Genome Project, rightfully hopeful spokespersons left the impression that genetic knowledge would be all-explanatory. Geneticist Walter Gilbert said that "to identify a relevant region of DNA, a gene, and then to clone and sequence it is now the underpinning of all biological science."[9] Nobel Prize winner James Watson told *Time*, "We used to think our fate was in our stars. Now we know, in large measure, our fate is in our genes."[10] The DNA sequence has been called the Rosetta Stone and the Holy Grail and even "the ultimate explanation of human being." This is the scientific seed that has grown into the gene myth.

The key belief in the gene myth is genetic determinism. "It's all in the genes!" we say. Genes make us tall or short. Genes determine our eye color. So-called "defective genes" are responsible for diseases such as cancer, Alzheimer's, Huntington's, Tay Sachs, and cystic fibrosis, to name a few. Some genes make us fat. Might there also exist genes that predispose us to sexual orientation? alcoholism? violence? Do genes govern our behavior? Does DNA answer the age-old question: Who am I?

In the decades following the Second World War the myth of environmental determinism became near dogma in Western culture. On the assumption that we are born a *tabula rasa*, a clean slate, upon which experience could write its own text, positive parental influence and quality of education became the preferred means for improving the human lot. As the optimism of the 1950s deteriorated gradually in the direction of postmodern pluralism, the West has come to see us as profoundly determined by our social location: by the cultural dictates of our language, class, race, and gender. As we approach the century's end we can see that the grip of environmental determinism has begun to loosen, however. A new conceptual canon is beginning to become dogma. This new canon is the gene myth. Gone is the clean slate. Now we are said to be born with much already written on that slate. Rather than see ourselves as determined by external environmental forces, we are gradually seeing ourselves as determined by internal biological predispositions.

The genetic determinism of the gene myth, curiously enough, has two distinct faces. The first is the fatalistic face. I call it *puppet determinism*. According to puppet determinism, the DNA defines who we are, and the genes, like a puppeteer, pull the strings that make us dance. To speak of "genetic essentialism" or to see genetics as "the ultimate explanation of human being" is to place DNA in the position of defining who we are and who we can be. To speak of "genetic fatalism" or to say "It's all in the genes" is to assume that genetic influences are unchangeable, that we are immutably destined to act as our DNA programs us to act. The psychological corollary to genetic determinism is clear: what we assume to be personal freedom is only an illusion. The ethical corollary to genetic determinism is also clear: we are not responsible for what we do; our genes are. We have natural innocence in a new form: blame my genes, not me.

The second is the future face. I call it *promethean determinism*. It is a version of what I have elsewhere identified as the understanding-decision-control (u–d–c) formula, characteristic of the modern doctrine of technological progress.[11] Promethean determinism assigns our scientists the task of *understanding* just how the genes work plus that of making the *decision* to develop appropriate technologies based upon this understanding; and this will then give the human race *control* over what nature has bequeathed to us. Because the history of genes constitutes the history of human evolution, once we have gained control we will be able to guide the future evolution of the human race. We will have wrested from nature her secrets, and this will transform us from the determined into the determiners.

Do these two faces of determinism look alike? Are they compatible? One significant difference is notable. In the case of puppet determinism, there is no human self that transcends our DNA. Genetic essentialism means that who we are is exhaustively determined by our genes. There is no "I" or "we" that stands independently. The self is a puppet. What we know as human freedom simply does not exist, except as an illusion.

Promethean determinism seems to entail the unspoken trust that some sort of decision-making entity, perhaps the human self, has gained a degree of critical distance from its own genetic make-up. This self is presumed without question to exist, and it can't be exhaustively reduced to its genetic determinants. The self is in charge, not the genes. The genes determine the human future, but the human race determines the genes.

Despite the difference outlined here, genetic determinism as we find it in the gene myth includes both. The myth is not itself a highly developed philosophy or ideology that requires internal consistency. Rather, it is a barely articulated thought form ever present in contemporary culture which provides a framework for interpreting the human reality in light of genetic research.

In chapter 2 we will look closely at the scientific sources for puppet determinism. I will show how the work of molecular biologists in the Human Genome Project along with twins studies by behavioral geneticists combined with the grand story of nature told by evolutionary psychologists all conspire to convey the one message: "It's all in the genes." Here, however, we will look a little more closely at promethean determinism, because in the countenance of this face we find simultaneously the smile of creativity and the frown of disapproval. The frown of disapproval has to do with the prohibition against playing God. But first, Prometheus.

Prometheus and the Gene Myth

The term *myth* as I use it in the phrase *gene myth* refers to a thought structure, a set of conceptual assumptions about reality that frames and filters the cultural reception of new scientific knowledge. Yet, there is a much more classic understanding of myth. In premodern society a myth was a story about the gods, a story that explains how the gods created the world or a part of the world in the beginning; and this story explains why things are the way they are today. To one of these classical myths we must now turn if we are to understand how the gene myth frames reality today. It is the myth of Prometheus.

Do you want to know how we came to have fire? The myth of Prometheus answers that question. When the world was being created, the sky god Zeus was in an ill mood. He decided to withhold fire from earth's inhabitants, leaving the human race to relentless cold and darkness. The farsighted maverick Titan Prometheus could see the potential value of fire for nascent humanity. He could foresee its value for warming homes and providing lamplight for reading good books late at night. He could see that it would be the gift of fire that would separate humanity from the beasts, making it possible to forge tools. So he craftily snuck up into the heavens where the gods dwell, and where the sun is kept. He lit his torch from the fires of the sun and then carried this precious source of heat and light back to earth.

Prometheus' name is interesting. With the prefix *pro*, meaning before or ahead, attached to *mathein*, the infinitive of the Greek verb to know or understand, his name means forethought or thinking ahead. Those knowing the myth also remember how Zeus, angry over Prometheus' impertinence, exacted merciless punishment on the rebel. Zeus chained the Titan to a rock where an eagle could feast all day long on his liver. The sky god cursed the future-oriented Prometheus by saying, "Forever shall the intolerable present grind you down." Although Prometheus is usually remembered as the example of human pride or *hubris* for having stormed the realm of heaven where "No Trespassing" signs keep most mortals out, we can also remember him for his name: forethought.

What happens to Epimetheus complements and reinforces the point of the Prometheus story. Seldom remembered but equally important is Prometheus' brother, Epimetheus. The prefix *epi* refers to what comes after, as in words such as "epilogue," so this brother's name means afterthought. Epimetheus was not the kind of person who would plan ahead. He failed to anticipate consequences. He lived for the moment, for the pleasures of the moment.

Zeus, still beside himself with anger over the pride of Prometheus, decided to add to his revenge. With the help of other Olympians, he fashioned a beautiful woman by the name of Pandora. He then sent the lovely woman into Epimetheus' neighborhood. Prometheus had earlier warned his brother to avoid any gifts sent from Zeus. But once Epimetheus' eyes fixed on the pulchritudinous Pandora, he could not wait to ask anyone for advice. He married her immediately and took her into his house.

Now, it was in the house of Epimetheus that the great box was kept, the box in which all the evil of the world had been locked so as to protect the human race. As long as the lid remained tightly shut, no evil could escape to harm anyone. But Pandora's curiosity was triggered, and she wondered what could be in that box. She opened the lid. Immediately there flew out a multitude of plagues: gout, rheumatism, envy, spite, hatred, revenge, and all forms of evil that spread far and wide over the earth. By the time Pandora could slam the lid back down on the box, only one thing remained. That was hope. All but hope had escaped to wreak havoc.

It is the combination of pride and afterthought that brings evil upon the human race. This is the message of the Prometheus myth. Yet, we might ask: What about forethought? We might further ask: What about the spirit of the Titan who rebelled against the arbitrariness of tyranny?

If the gene myth has a dark side, we might at first suspect we would find it in the shadows of puppet determinism. After all, DNA, our creator, still keeps us in the cold dark realm of disease and suffering. Like a political tyrant, DNA gives orders and we take them. What would happen if we, like Prometheus, refused to accept our DNA destiny? What if we rebel? What if we strike out in behalf of liberation? What if we sneak into the realm of the tyrant, steal its secret, and then use the genetic secret to enhance the quality of human life? Would we be storming the citadel of Zeus? Would we be guilty of excessive human pride? Would we be playing God?

It is in promethean determinism that contemporary critics of genetic science claim to find the dark side. The gene myth comes complete with taboos. One of the taboos is this: don't play God!

What needs to happen, I think, is for us to gain a clearer understanding of just how the gene myth frames and filters the way we in our culture interpret the new genetics. In doing so, we may find that the gene myth misleads us in some important respects. But if it is in fact the case that the gene myth misleads, we will see its deception only if we focus on the frame for a while. Rather than look squarely at genetics, we need to pause to examine the way we structure our understanding of genetics. One element in this structure is the taboo against playing God. To the question of playing God we now turn.

What Does "Playing God" Mean?

To ask whether we are asking our scientists to play God implies a set of correlate questions. Does our genetic make-up represent a divine creation in such a way that it is complete and final as it is? Is our DNA sacred? Are we desecrating a sacred realm when we try to discern the mysteries of DNA? Are we exhibiting promethean pride when we try to engineer our genetic future? Are we risking epimethean afterthought and opening Pandora's box?

The enigmatic phrase "playing God" can have at least three overlapping meanings. The first and somewhat benign meaning has to do with *learning God's awesome secrets*. It refers to the sense of awe rising up from new discoveries into the depths of life. Science and its accompanying technology are shedding light down into the hitherto dark and secretive caverns of human reality. Mysteries are being revealed; and we the revealers sense that we are on the threshold of acquiring "Godlike" powers. At this level we do not yet have any reason to object to research. Rather, what we have here is an expression of awe.[12] Bioethicist Thomas A. Shannon writes, "We are genuinely on the edge of a new revolution in medicine, one that will provide access to the very structure of our nature. We can literally reach inside ourselves, remove a gene or genes, and either correct or replace them. Such power is truly awe-inspiring."[13]

The second meaning of "playing God" has to do with the actual wielding of *power over life and death*. This applies, for example, to medical doctors working in the clinical setting with an emergency surgery. The patient feels helpless. Only the attention and skill of the surgeon stands between the patient and death. The doctor is the only door to life. The patient is utterly dependent upon the physician for his or her very existence. Regardless of whether or not doctors feel they have omnipotence in this situation, the patients impute it to them.[14]

The medical meaning of "playing God" operates with two assumptions. First, decisions regarding life and death belong to God's prerogative and not to human beings. Second, when we humans make life-and-death decisions we exhibit *hubris* or pride—that is, we overreach ourselves and transgress divinely imposed limits. These assumptions create anxiety; and this anxiety energizes and drives endless debates in medical ethics over when to pull the plug, or even over whether we should pull the plug at all.

Some people would like to challenge these two assumptions. Stanford medical ethicist Ernlé W.D. Young, for example, complains that the charge of playing God seems to be leveled only at doctors. This is unfair, he says. Architects and contractors construct bridges that on occasion collapse and kill people, yet no one charges them with playing God. Military commanders are not criticized for playing God when their decisions lead to the death of soldiers in combat. And we ourselves, when deciding to continue smoking or eating unhealthy foods, are in fact making death decisions while enjoying exemption from the playing-God charge. Ernlé Young raises this complaint in defense of the medical profession. Doctors frequently find themselves in situations where they cannot simply let nature take its course, where they must make the hard decisions. Doctors in this situation should be supported ethically, he believes; they should not be accused of trespassing on God's property. Physicians, like the rest of us, are endowed with the gift and burden of freedom, and this means we cannot escape the responsibility of shouldering decision making in difficult circumstances. Young is unconvinced that the category "playing God" should exist at all.[15]

A third meaning of "playing God" is the one that concerns us here, namely, the use of science to *alter life and influence human evolution.* "Playing God" in this case means that we—at least the scientists among us—are substituting ourselves for God in determining what human nature will be. It refers to placing ourselves where God and only God belongs.

The power to alter human nature as a life form similarly evokes the question: Should only God be doing this? Fifty-eight percent of those who answered a *Time*/CNN poll said they think altering human genes is against the will of God.[16] In a 1980 letter of warning to then President Jimmy Carter, several Roman Catholic, Protestant, and Jewish spokespersons used the phrase "playing God" to refer to individuals or groups whom they alleged were seeking to control life forms. The letter writers insisted that any attempt to "correct" our mental and social structures by genetic means to fit one group's vision of humanity is dangerous.[17] The problem with genetic engineering and other reproductive technologies, says Pope John Paul II, is that they place our destiny in our own hands and lead to the "temptation to go beyond the limits of a reasonable dominion over nature."[18]

Why do such critics of genetic research prescribe a new commandment, "Thou shalt not play God"? The answer here is this: because promethean pride or *hubris* is dangerous.[19] We have learned from experience that what the Bible says is true: "Pride goes before destruction" (Proverbs 16:18). And in our modern era, pride among the natural scientists has taken the form of overestimating our knowledge, of arrogating for science a kind of omniscience that we do not in fact have. Or, to refine it a bit: "playing God" means we confuse the knowledge we do have with the wisdom to decide how to use it. Frequently lacking this wisdom we falsely assume we possess, scientific knowledge leads to unforeseen consequences such as the destruction of the ecosphere.[20] This is the genetic version of the Frankenstein legend made into a cultural icon through Michael Crichton's novel *Jurassic Park*.[21]

Although the phrase "playing God" is foreign to theologians and is not likely to appear in a theological glossary, some religious spokespersons employ the idea when referring to genetics. A National Council of Churches task-force statement recognizes that "human beings have an ability to do Godlike things: to exercise creativity, to direct and redirect processes of nature. But the warnings also imply that these powers may be used rashly, that it may be better for people to remember that they are creatures and not gods."[22] The theology that tells us to remain within limits is forcefully stated by the 1992 United Methodist Church statement on genetics: "The image of God, in which humanity is created, confers both power and responsibility to use power as God does: neither by coercion nor tyranny, but by love. Failure to accept limits by rejecting or ignoring accountability to God and interdependency with the whole of creation is the essence of sin."[23]

Applied to genetic therapy, the commandment against "playing God" by transgressing limits implies that the unpredictability of destructive effects on the human gene pool should lead to a proscription against modifying the germline—that is, against engineering those genes that will be passed on to future generations. In light of this, "there is general agreement that human germ-line intervention for any purpose should always be governed by stringent criteria for safety and predictability."[24] This "general agreement" seeks to draw upon wisdom to mitigate pride.

A correlate to this third meaning of "playing God" is that DNA has come to function in effect as an inviolable sacred entity, a special province of the divine that should be off limits to mere mortals. Geneticist Robert Sinsheimer, among others, suggests that when we see ourselves as the creators of life, we lose reverence for life.[25] It is just this lack of reverence for life as nature has bequeathed it to us that drives political activist Jeremy Rifkin to attack the kind of genetic research that will lead to algeny—that is, to "the upgrading of existing organisms and the design of wholly new ones with the intent of 'perfecting' their performance." The problem with algeny is that it represents excessive human pride. "It is humanity's attempt to give metaphysical meaning to its emerging technological relationship with nature."[26] The message here is this: let nature be! Don't try to make it better! In advocating this hands-off policy, Rifkin does not appeal to Christian or Jewish or other theological principles. Rather, he appeals to a vague naturalism, according to which nature itself claims sacred status. He issues his own missionary's call: "The resacralization of nature stands before us as the great mission of the coming age."[27]

This position has garnered critics. Political scientist Walter Truett Anderson thinks Rifkin's attack against genetic engineering is an unnecessarily hysterical one, and he dubs it "biological McCarthyism." Anderson's own position is promethean. He believes that the human race should become deliberate about the future of its own evolution. "This is the project of the coming era: to create a social and political order—a global one—commensurate to human power in nature. The project requires a shift from evolutionary meddling to evolutionary governance, informed by an ethic of responsibility—an evolutionary ethic, not merely an environmental ethic—and it requires appropriate ways of thinking about new issues and making decisions."[28]

Why treat nature in general, or DNA specifically, as sacred and therefore morally immune from technological intervention? Ronald Cole-Turner, a theologian and ethicist who specializes in bioethics, criticizes Sinsheimer and Rifkin for making an unwarranted philosophical and theological leap from the association of DNA with life to the metaphysical proscription against technical manipulation. "Is DNA the essence of life? Is it any more arrogant or sacrilegious to cut DNA than to cut living tissue, as in surgery? It is hard to imagine a scientific or philosophical argument that would successfully support the metaphysical or moral uniqueness of DNA. Even DNA's capacity to replicate does not elevate this molecule to a higher metaphysical or moral level. Replication and sexual reproduction are important capacities, crucial in biology. But they are hardly the stuff of sanctity."[29]

To nominate DNA for election into the halls of functional sacredness, says Cole-Turner, is arbitrary. Theologians in particular should avoid this pitfall. "To think of genetic material as the exclusive realm of divine grace and creativity is to reduce God to the level of restriction enzymes, viruses, and sexual reproduction. Treating DNA as matter—complicated, awe-inspiring, and elaborately coded, but matter nonetheless—is not in itself sacrilegious."[30] There is something special about DNA, to be sure; the particular genome that DNA bequeaths to each of us is largely determinative of our individual identity. Yet this is insufficient reason for treating it as functionally sacred.

What Is Sacred: God or the Creation?

What the three meanings of "playing God" raise up for us is the question of the relationship between the divine creator and the natural creation. Theists in the Jewish and Christian traditions are clear: natural life, important as it is, is not ultimate. God is ultimate.

One can argue to this position on the basis of *creatio ex nihilo*, creation out of nothing. All that exists has been called from nothing by the voice of God and brought into existence. Life, as everything else in existence, is finite, temporal, and mortal. The natural world depends upon a divine creator who transcends it. Nature is not its own author. Nor can it claim ultimacy, sanctity, or any other status rivaling God. This leads biologist Hessel Bouma III and his colleagues at the Calvin Center for Christian Scholarship to a pithy proposition: "God is the creator. Therefore, nothing that God made is god, and all that God made is good." This implies, among other things, that we should be careful when accusing physicians and scientists of "playing God." We must avoid idolatrous expectations of technology, to be sure; "but to presume that human technological intervention violates God's rule is to worship Mother Nature, not the creator. Natural processes are not sacrosanct."[31]

One can also argue to this position on the basis of *creatio continua*, continuous creation. On the basis of the idea that creation is ongoing, one can argue for human intervention and contribution to the process. God did not just extricate the world from the divine assembly line like a car, fill its tank with gas, and then let it drive itself down the highways of history. Divine steering, braking, and accelerating still go on. The creative act whereby God brought the world into existence *ab initio*, at the beginning, is complemented with God's continued exercise of creative power through the course of natural and human history. The God of the Bible is by no means absent. This God enters the course of events, makes promises, and then fulfills them. God is the source of the new. Just as the world appeared new at the beginning, God continues to impart the new to the world and promises a yet outstanding new creation still to come.

My own way of conceiving of *creatio ex nihilo* together with *creatio continua* is this way: the first thing God did was to give the world a future.[32] The act of drawing the world into existence from nothing is the act of giving the world a future. As long we have a future, we exist. When we lose our future, we cease to exist. God continues moment to moment to bestow futurity, and this establishes continuity while opening reality up to newness. Future-giving is the way in which God is creative. It is also the way God redeems. God's grace comes to the creation through creative and redemptive future-giving.

God creates new things. The biblical description of divine activity in the world includes promises and fulfillments of promises. This implies two divine qualities. First, God is not restricted to the old, not confined by the *status quo*. God may promise new realities and then bring them to pass. The most important of the still outstanding divine promises is that of the "new creation" yet to come. Second, this God is faithful, trustworthy. On the basis of the past record, the God of Israel can be trusted to keep a promise. For us this means that we can trust God's creative and redemptive activity to continue in the future.

The next step in the argument is to conceive of the human being as the created cocreator. The term *created cocreator* comes from the work of Philip Hefner.[33] This term is important for at least two reasons. First, the term "created" reminds us that God creates differently from the way we human beings create. God creates *ex nihilo*. We have been created by God. We are creatures. So, whatever creativity we manifest cannot rank on the same level as creation out of nothing, on the same level with our creator. Yet, secondly, the term *cocreator* signifies what we all know, namely, the creation does not stand still. It moves. It changes. So do we. And, furthermore, we have partial influence on the direction in which it moves and the kind of changes that take place. We are creative in the transformative sense. Might we then think of the *imago dei*—the image of God embedded in the human race—in terms of creativity? Might we think of ourselves as cocreators, sharing in the transforming work of God's ongoing creation?

Human creativity is ambiguous. We are condemned to be creative. We cannot avoid it. The human being is a toolmaker and a tool user. We are *homo faber*. We cannot be human without being technological, and technology changes things for good or ill. Technology is normally designed for good reasons such as service to human health and welfare, but we know all too well how epimethean shortsightedness in technological advance does damage. This is indirect evil. Direct evil is also possible. Technology can be pressed into the service of violence and war, as in the making of weapons. Technology is by no means an unmitigated good. Despite its occasional deleterious consequences, we humans have no choice but to continue to express ourselves technologically and, hence, creatively.

We cannot not be creative. Our ethical mandate, then, has to do with the purposes toward which our creativity is directed and the degree of zeal with which we approach our creative tasks.

At the level of culture and public policy, the gene myth is likely to exert considerable influence on our creative zeal. Negatively, the myth's puppet determinism could retard enthusiasm by introducing fatalism. Why put forth effort if our fate is already determined by the genes? If it's all in the genes, should we not simply accept the *status quo?* Combined with the ascription of sacredness to DNA and its taboo against playing God, such a genetic fatalism could lead to public policies that limit, if not retard, scientific research. Positively, however, the mood of promethean heroism will continue to inspire zeal for scientific and technological creativity. The double-sidedness of promethean determinism—heroic creativity balanced with caution against playing God—embodies the ambivalence of hope and trepidation precipitated in our society by the genetic revolution.

This ambivalence of hope and caution animates our discussion here. In particular, the confusion created by the gene myth regarding human freedom and ethical responsibility warrants attention. Just what kind of freedom appears compromised or eliminated by genetic determinism? Does genetic determinism necessarily contradict our understanding of human freedom, or might it actually enhance it? And, are there any implications for the concept of Christian freedom?

Four Freedoms

Puppet determinism seems to take away freedom we thought we had, but promethean determinism seems to enhance our freedom to influence the future. Is this a contradiction? Is this a paradox? Just what kind of freedom is affected by the rising sense of genetic determinism?

At least four types of human freedom are worth sorting out here: political freedom known as liberty; natural freedom understood as freedom of the will; moral freedom or the capacity to pursue what is good; and future freedom associated with creativity. The second and fourth types of freedom are most directly affected by talk of genetic determinism.

First, *political freedom* or liberty. This is social freedom understood as independence from external coercion or constraint. External circumstances vary from context to context, some permitting greater individual autonomy and others depriving us of freedom. This circumstantial freedom permits us to do what we please, to carry out in overt action decisions we have reached. Liberty is granted us by the *polis*, given by the social situation within which we find ourselves. It is external. It is freedom granted us regardless of our internal dispositions, desires, or decisions. Whether motivated to do good or to do evil, circumstantial freedom permits us to express our will, to act according to our own individual motives.

The gene myth has little to say directly about political freedom, because the field of genetics turns our attention away from external and toward internal determinism. Rather than environmental circumstances that might coerce us from the outside, genetic determinism looks for internal coercion.

Indirectly, however, the gene myth may say quite a bit about political philosophy. At least the critics think so. Critics of the gene myth react with horror, fearing that the doctrine of genetic determinism provides scientific support for conservative social theory and practice. Such critics typically pay no attention to promethean determinism, however. Genetic determinism in its promethean form is future oriented and embraces the kind of social change that attracts most people in a liberal society.

Second, *natural freedom* or freedom of the will. What we experience as free will is traditionally assumed to be inherent in human nature. It comes as a birthright. It comes to expression as choice, as free choice, as the ability to choose before we act. It is the freedom of the individual human being as an agent who acts. Beyond determination by external circumstances, each of us possesses an innate power to render value judgments, to make decisions, and to take actions that affect our external environment and our internal character.

When speaking of this free will, I believe we should be careful to avoid reifying the will. The will is not a thing. It is not an item that takes up time and space while engaging in choosing among alternatives. Rather, the locus of freedom is the human self or person; and we use the word "will" to refer to the ability of the self or person to make choices. That we are born with a dimension of personhood seems obvious to our parents. They can see our sense of self come to expression and take shape over years of development. Nature seems to endow each of us with the capacity for centered activity, for self-orientation, for self-initiation, and for decision making.

The concept of free will, no matter how intuitively obvious it seems to be, has in recent decades been challenged by one or another form of deterministic thinking. Environmental determinism has dominated until the more recent rise of the gene myth. Whether by environmental nurture or inborn nature, determinists suggest that they can explain away what we experience as free will. It is freedom understood as free will that seems to be most under attack by the puppet determinism of the gene myth. This is because DNA is thought to function as an internal puppeteer pulling the strings that determine our decisions and actions. According to the gene myth, the restrictions on our liberty do not come merely from outside, from the political or physical environment; rather, they come from inside, from our DNA as well. We now picture ourselves as trapped between two tyrants, external environment and internal DNA, with DNA the stronger. In the picture drawn by the gene myth no person or self can be seen, only the tug of war between genes and environment. Instead of nature bequeathing us free will, it gives us at best only the delusion of free will. The task of the science of genetics, so the gene myth says, is to expose the delusion.

Third, *moral freedom*, which refers to the life of wisdom or virtue. It consists in the capacity to pursue the good. The capacity to pursue the good is not fully given us by our inborn nature. Rather, it is an acquired freedom. When a person pursues goodness in life he or she develops moral character or virtue—that is, the habitual disposition to desire what is good. The desire for the good is manifested in choosing what is right, true, and helpful. Some of the obstacles or hurdles to be jumped when pursuing what is good stem from internal appetites or passions that generate desires and mental attitudes which conflict with what is good, tempting us to make wrong rather than right choices. The ancient philosophers described these temptations coming from within us as exacting bondage on our will, as a form of internal constraint or even compulsion.[34] Stoic philosopher Epictetus said that only a person who had acquired virtue could be truly free; an evil person remains in bondage.

Moral freedom entails an orientation of the self toward the good. This orientation toward the good is not just an accessory to the self, as if the self could treat the question of doing good or evil as merely optional. Rather, choosing in behalf of the good is built into the very development of the self itself. Choosing for the good is a freedom enhancer, so to speak. I find the work of philosopher Charles Taylor instructive here. "Selfhood and the good, or in another way, selfhood and morality, turn out to be inextricably intertwined themes."[35] In certain respects, it is the orientation toward what is good that signals the transcendence of the person beyond his or her biological substrate. Taylor again:

> A self or a person ... is not like an object in the usually understood sense. We are not selves in the way that we are organisms, or we don't have selves in the way we have hearts and livers. We are living beings with these organs quite independently of our self-understandings or -interpretations, or the meanings things have for us. But we are only selves insofar as we move in a certain space of questions, as we seek and find an orientation to the good.[36]

In order for the self to become a true self, it needs to be oriented beyond itself toward what is good. The acquiring of moral character is a form of liberation, then, a freeing of the will from self-orientation in order to pursue what is truly good. The Christian variant on this theme is to define the good in terms of *caritas* (charity), as love for God and love for neighbor. The freedom to love truly is a freedom that God gives us. This freedom is a gift of grace. On this point the Christian view differs from that of the classics, especially the Stoics. Whereas the Stoics engaged in active efforts to acquire freedom by gaining dominion over oneself, the Christians announce that freedom first comes to us as an act of God's grace, as an event of liberation. We will look again at the concept of Christian freedom later.

With regard to the gene myth, moral freedom is superficially treated as a subspecies of free will and given an explanation in terms of both types of genetic determinism. According to puppet determinism, human morality, including altruism, is a product of the selfish gene traveling its way up the road of evolutionary development. According to promethean determinism, we in the human race are now expanding our powers to determine what is good in ever more complex fashion. We will examine this component to the gene myth in greater detail in chapter 2.

Fourth, *future freedom* has to do with human creativity. We human beings have inherited from nature memory, imaginations, and free will. These three together encourage us to imagine a future that will be different from the past and present. The deliberations and decisions and actions we take are frequently motivated by the desire to create something new, to alter our environment or our personal character according to a vision of a better reality. Whether engineering and building a new machine or making a New Year's resolution, we take action intended to give direction to the future course of events. Future freedom consists in transcending at least to some degree the determinism of the past, making ourselves into a determinant for what will happen in the future.

It is future freedom that is assumed and capitalized on by the gene myth with its promethean determinism. The human capacity for influencing the future becomes the capacity for influencing what will be human in the future. In a way analogous to the acquired freedom of virtue, the gene myth proffers a self-referential determinism. Rather than deal with an individual person striving to transcend appetites and passions by developing moral character, the gene myth deals with the human race as a whole. The vague assumption made by the gene myth is that humanity as a whole can employ technology to alter its appetites and passions. To influence our own evolutionary development is to invoke human creativity, to exercise the freedom that makes self-determination possible—in this case, the human race as a totality is the self doing the determining.

Freedom and Beneficence

As we have seen, theologians have a particular stake in moral freedom, especially the variant known as Christian freedom. Although moral freedom is frequently conflated with freedom of the will, some differences are worth noting. Freedom of the will is normally associated with natural freedom, something bequeathed to us by nature. It appears that we are born with free wills, as our parents can attest. We might eventually even say that freedom of the will is genetic, inherited from our evolutionary history. The salient characteristic of the free will is that it—or better, the person with the free will—chooses among alternatives. Thomas Aquinas defined what we call natural freedom as the power of the will to choose between alternatives that are related as means to ends—that is, to choose according to one's desires and designs.[37] The key question here is this: Who defines our desires or beliefs? When they are defined by us, then the will is bound by the compulsions of the self. When they are defined by God, then the will is freed from the compulsions of the self. Even though the human will is by definition free to choose, the theological question as to whether it is bound or free lies at a level deeper than choice.

Moral freedom, therefore, is acquired. It is not innate. Yet we might still ask: Is it genetic? Is it the result of genetic and environmental factors in tandem? Does it require a divine intervention over and above genes and environment? Arriving at answers to such questions will not be easy.

The variant of moral freedom to which we now turn, Christian freedom, takes the form of liberation. It is liberation from sin. It is not simply liberation from nature, because nature, theologically understood as the product of God's creative work, actually inclines us toward the good. Nature perceived scientifically hides this truth, as we will see later when discussing the selfish gene and natural selection. To see the lure of nature toward what is good requires discernment; it requires a vision of divinely promised redemption. Scientifically interpreted, nature is neutral; it is neither good nor bad. Theologically interpreted, nature is essentially good. Is our nature only biological nature?

Nature as we daily experience it is ambiguous, fraught with benefits and liabilities. On a daily basis we experience ourselves and the world around us as both beautiful and fallen, as tarnished or contaminated with injustice and suffering. We wake up as infants or toddlers and find ourselves subject externally to both love and abuse. We also find ourselves driven from within by compulsive desires and forces that seem to have been put there by nature without asking our permission. Awakening to self-consciousness we find our will already at war within us, as biological needs and desires contend with one another in our decision making. Gaining self-control becomes a daily battle, some days showing victory and others defeat. And the paradox is this: these biological determinants come from within, not just without. So, we pause to ask: Are they me? Am I my biology? If so, then why do I sometimes feel victimized by biological compulsions?

In part, what theologians call sin consists of yielding to inner biological compulsions. But by no means is sin limited to this. Perhaps it is more appropriate to say that our human will, as it has been determined by our biological inheritance, is originally oriented toward the dictates of that biology. In establishing its relative independence from biology, the human self or person utilizes the power of the will. But the process of writing its own declaration of independence results in what is inevitable, namely, the establishment of a self that is oriented toward itself. This self-orientation, manifested both as individual selfishness and as group self-interest, from Cain's murder of Abel down to the mass genocides of the twentieth century, is the source of human-wrought evil in the world. It is this orientation of the self toward itself from which Christian freedom seeks liberation.

When confronting this paradoxical struggle, the wise philosophers of ancient Greece as well as the seers of Upanishadic India and other mystics drew a sometimes crude line between the physical and the spiritual. A state of war exists between body and soul, they contended, a microcosm of the macrocosmic battle between matter and spirit. In order to gain release from the chains of the physical body, mystics developed meditation techniques and philosophers lifted our minds to envision rational virtues such as truth, beauty, and justice. Although cast as a struggle between body and soul, in some ways the spiritual quest became a search for the experience of transcending the limits of the self, at least the self as constrained by its biological basis. What we know as moral freedom is acquired through the study of philosophy or through engaging in spiritual discipline—that is, pursuing our so-called higher nature.

The New Testament church inherited this already established cultural construct. The classic dualistic interpretation of the human predicament framed Christian anthropology just as the gene myth frames our interpretation of science today. Although on the surface it appears that Christian theologians simply joined the spiritual army in its attack against physical forces, a closer look will show that the real battlefield was elsewhere. Christians had no stake in supporting a spiritual defeat of the body. After all, God had become physical in the incarnation. The incarnation took place because God "so loved the world" (John 3:16). Nature, even biological nature, must in some sense be deemed good if God would become so invested in it.

If the real battlefield was not where body contends with spirit, then where was it? Between self and God. Between the self and its own transcendence. And, despite the best strategies of the wisest and noblest pursuers of virtue, the struggle for self-transcendence was destined for loss. The reason: the initiative for self-transcendence always begins and ends with the self. Even truth, beauty, and justice—as sublime and universal as they appear—are always subject to subversion and perversion by self-interest. The self cannot by itself divest itself of itself. And when the problem of sin is located in the orientation of the self toward itself, we have a problem indeed.

The concept of moral freedom as acquired freedom is certainly the right category, from the Christian point of view. But how do we get there from here?

The theological answer is divine grace. At the cosmic level, the reconciliation of the physical and spiritual is accomplished through the incarnation, the shared oneness of the divine spirit with the material world. At the personal level, God's Holy Spirit becomes present and empowers the self to transcend itself through love. "Freedom from sin results from charity," writes Aquinas, "which is poured forth in our hearts by the Holy Spirit, who is given to us (Rom. 5:5)."[38]

The Holy Spirit orients our self to delight in God, to desire what is good, and to love our neighbors. At least this is the way Augustine sees it. "Free choice alone, if the way of truth is hidden, avails for nothing but sin; and when the right action and the true aim has begun to appear clearly, there is still no doing, no devotion, no good life, unless it be also delighted in and loved." How does this delight and love become present in us? By God's grace present in the Holy Spirit. "The human will is divinely assisted to do the right in such manner that, besides man's creation with the endowment of freedom to choose, and besides the teaching by which he is instructed how he ought to live, he receives the Holy Spirit, whereby there arises in his soul the delight in and love of God, the supreme and changeless good. This gift is here and now, while he walks by faith, not yet by sight."[39]

Perhaps Martin Luther describes it in its starkest and most paradoxical form. In the opening paragraphs of his 1519 essay "The Freedom of A Christian," he states it bluntly:

A Christian is a perfectly free lord of all, subject to none.
A Christian is a perfectly dutiful servant of all, subject to all.[40]

Accepting as his conceptual set the classic dualism of spiritual and bodily natures, Luther stresses that when the inner life of the soul is solidly grounded in God, then nothing in the outer life of the body can enhance or harm it. Even if our body is dying, we can be rising to new life in the spirit. This experienced unity with God gives a person unassailable courage and self-confidence. Whether politically free or serving a prison term, whether in perfect health or suffering from a terminal illness, we can be "perfectly free" lords, "subject to none."

What grounds the inner person in God? Faith in Jesus Christ. In faith the Holy Spirit makes the resurrected and living Christ present to the believer. The presence of Christ effects an exchange of sorts, a unity of the divine and human in which their respective qualities get swapped. Luther uses the analogy of a marriage in which the respective properties become shared. Faith "unites the soul with Christ as a bride is united with her bridegroom," he writes. "By this mystery . . . Christ and the soul become one flesh. And if they are one flesh and there is between them a true marriage . . . it follows that everything they have they hold in common, the good as well as the evil." He then proceeds to assert that in Christ, God takes on sin and death in order to bestow on us grace and life. "Christ is full of grace, life, and salvation. The soul is full of sins, death, and damnation. Now let faith come between them and sins, death, and damnation will be Christ's, while grace, life, and salvation will be the soul's."[41]

The event of faith becomes the event of liberation for the human soul. The divine nature of Christ becomes transferred, so to speak, to the person of faith. This event of grace provides the foundation for a life of joyous freedom, the sense that one is a perfectly free lord, subject to nobody.

Yet the nature of Christ has an identifiable character, namely, self-giving love or charity. To live inextricably close to Christ is to absorb and emit this love. Hence, the second side of the paradox of freedom: the person of faith is a "perfectly dutiful servant of all, subject to all." The "all" to which we are subject here is the neighbor who has a need that we can meet. To love is to serve. To be a lord is to be a servant. Radical independence becomes voluntary interdependence. Freedom in this instance consists in freedom from a self that otherwise could only serve itself. In no sense does this imply an ontological dissolution of the self into a nonself, however. Rather, to put the matter in an ethical category, faith requires a strong independent self that freely takes someone else's agenda as its own.

The orientation of one's life becomes the need of the neighbor, not the need of the Christian to do good deeds or generally to be nice to people. We should be guided in all our works "by this thought and contemplate this one thing alone," says Luther, that we may "serve and benefit others" and consider "nothing except the need and the advantage" of our neighbor.[42] Freedom expresses itself as neighbor love (*die Nächstenliebe*). St. Paul describes this as "faith working through love" (Gal. 5:6).

Christian freedom is the root that has sprouted in the twentieth century as liberation theology and some forms of feminist theology. "Feminist theology today is, by definition, *liberation theology*, because it is concerned with the liberation of all people to become full participants in human society," writes Yale theologian Letty Russell.[43] Once liberated, what should we do? Love our neighbors. "For Christian women the experience of new freedom leads to new responsibility ... *being set free for service to others.*"[44]

Critics of liberation theology say that this school of thought has merely given Christian baptism to Enlightenment liberalism, merely blessed the Western idea of political liberty. Political liberty protects natural freedom by permitting us to choose publicly more and more of what we desire, to be sure; but, critics argue, the freedom to choose what we desire ought not be confused with Christian freedom. The critics are correct in distinguishing these different types of freedom. However, what the critics miss, I think, is the root motive out of which the zeal for political liberation grows. The root motive is neighbor love expressed as the zeal for justice. Christians devoted to the liberation cause feel empathy for people suffering from the combination of political oppression and poverty. They are driven to extreme measures not out of simple adherence to Western liberal political theory but out of a passionate sense of solidarity with the victims of injustice. Entering voluntarily into solidarity with victims is neighbor love.

As we turn to the field of genetic science and its increasing impact on society, what form will neighbor love take? Many forms, perhaps. But one immediately looms into our vision. It is the principle bioethicists know as *beneficence*. Beneficence means acting with *caritas*, charity. The biomedical principle is this: "We have a duty to help others further their interests."[45] More than merely avoiding doing harm, more than mere nonmaleficence, beneficence seeks to create a positive imbalance of good over harm. It seeks to enhance human flourishing. In the medical setting it means that physicians and nurses seek to alleviate the pain, suffering, and disability caused by disease or injury.[46] Using a duty or obligation approach to ethics, this principle says that "we should always act in ways that promote the welfare of other people."[47]

Christians have no patent on beneficence, of course. It is an ethical principle found in other religious traditions and supportable by secular ethics.[48] What is significant in the present context is that an expanded form of the beneficence principle might be considered an authentic expression of neighbor love. Applied to the broad field of genetic science, the pursuit of such research would be deemed a positive good when it is oriented toward the goal of relieving human suffering and improving the welfare of the human race. Going beyond therapy for specific individuals in the hospital setting, beneficence becomes an abstract principle for guiding policy regarding the purposes and ends of laboratory research. Beneficence is not playing God in the sense of substituting the human for the divine. Yet, it is living in a godly way. It is the exercise of moral freedom.

Ontology and Ethics

We are now looking at our subject matter in two different, yet complementary, ways. On the one hand, we are asking about the ontology of human freedom—that is, we are asking how freedom is grounded in the nature of human being in the world and in relation to God. If the doctrine of puppet determinism presupposed by the gene myth could be confirmed scientifically, would we then conclude that what we experience as freedom is in fact an illusion? Our daily experience is structured around each of ourselves as a person, a person who is a self. But, we now ask, what is the ontological status of that self? Can we reduce what we experience as a freely choosing self to single-factor determinism (by genes alone) or two-factor determinism (by genes plus environment); or can we legitimately speak of three-factor determinism consisting of genes, environment, and the self? And then, what about Christian freedom? It appears that this kind of inner freedom depends upon a relationship to God that to some degree transcends our biology and transcends even the self-orientation of the self to itself. Or will this notion turn out to be a delusion as well?

On the other hand, we are asking about the ethics of genetic research and the development of gene-based technologies. Should such research go forward? Certainly the neighbor-love component to Christian freedom would seem to affirm that, yes indeed, such research should proceed apace and be pressed into the service of beneficence. Such an ethical affirmation is predicated on an ontological affirmation regarding God's grace and human freedom. Yet, we might ask: Do we run the risk here of begging the question? Do we run the risk of assuming freedom and neighbor responsibility just when this is being questioned by the gene myth? If puppet determinism holds and what we thought was human freedom turns out to be a sham, have we lost the specifically Christian grounding for supporting the advance of genetic research?

But what about promethean determinism? This brand of determinism, also at home in the gene myth, silently presupposes a self that is sufficiently independent to be in a position to envision ends and to place scientific research and technological development into the service of those ends. Promethean determinism does not stand in opposition to human freedom. It itself presupposes a form of human freedom, namely, future freedom. Curiously, promethean determinism stands in partial conflict with another doctrine within the gene myth, namely, the proscription against playing God.

Christian freedom and promethean determinism have something in common, namely, future freedom. Both begin with a vision of a better future, and both affirm that human action today can have positive consequences tomorrow. Yet, there are differences. The problem with prometheanism is that it pits itself against the divine. Prometheanism tries to play God by taking God's place, by taking control. It bears a mood of elimination of God and domination over nature rather than cooperation with God and nature. This mood of domination does not fit those who understand the immense complexities of molecular processes and are humbled before them. Research scientists are struck with a kind of awe and appreciation that makes domination and control seem unrealistic.

Despite the fact that domination and control are as morally undesirable as they are scientifically unrealistic, there is some truth to the belief that some things can be done and perhaps should be done to influence the course of our genetic future. Such things may be quite modest on a grand evolutionary scale, yet they can have an immense impact on the quality of life for certain individuals. Science in the service of beneficence ought not to be intimidated by a "No Tresspassing" sign that says, "Thou shalt not play God." Rather, science in the service of beneficence means we are playing human in a free and responsible way.

< TWO >

Puppet Determinism and Promethean Determinism

Solved by standard Gammas, unvarying Deltas, uniform Epsilons.
Millions of identical twins. The principle of mass production at last
applied to biology.

—Aldous Huxley, *Brave New World*

We are witnessing the end of the scientific era. Science, like other
outmoded systems, is destroying itself. As it gains in power, it proves
itself incapable of handling the power.

—Michael Crichton, *Jurassic Park*

Perhaps we could describe our belief in genetic determinism as a three-legged stool. The round seat on the stool's top represents the gene myth—that is, our cultural belief in genetic determinism of both the puppet and the promethean types. Supporting the seat are three apparently scientific legs: molecular biology, behavioral genetics, and evolutionary psychology.

The gene myth has its friends and enemies. Its friends include molecular biologists, behavioral geneticists, and evolutionary psychologists—also known as sociobiologists—who emphasize genetic determinism for its shock value and to dramatize the importance of their work.

Yet, some natural scientists and social scientists refuse to sit on the stool of genetic determinism. Harvard geneticist Richard Lewontin along with Steven Rose and Leon Kamin, for example, are uncomfortable with genetic determinism because it appears to lead to conservative political ideology. The determinism of the gene myth tends to support right-wing rather than left-wing social philosophies. Lewontin along with his colleagues describe the view they oppose:

> [*Biological determinism* holds] that human lives and actions are inevitable consequences of the biochemical properties of the cells that make up the individual; and these characteristics are in turn uniquely determined by the constituents of the genes possessed by each individual. Ultimately, all human behavior—hence all human society—is governed by a chain of determinants that runs from the gene to the individual to the sum of the behaviors of all individuals. The determinists would have it, then, that human nature is fixed by our genes.[1]

What would Lewontin put in the place of biological determinism? Free will? No. He advocates two-factor determinism, genes plus environment. "We must insist that a full understanding of the human condition demands an integration of the biological and the social in which neither is given primacy or ontological priority over the other."[2] He is not replacing genetic determinism with social determinism, just adding the latter to the former. He combines the two. What Lewontin wants to avoid is a social philosophy that says: if human nature is fixed by the genes, then we cannot change society. Instead, he rallies us for social change. He advocates social reform. And he wants a science that will support social advance. "'Natural' is not 'fixed.' Nature can be changed according to nature."[3]

Significant for our study here is the fact that most critics of genetic determinism are also determinists. They simply exchange one-factor determinism, genes alone, for two-factor determinism, genes plus environment. Although the concept of freedom will pop up within the discussion, it is seldom integrated into a scientific understanding of human nature.

It is also significant to repeat that a glaring inconsistency persists in the operative concept of genetic determinism. On the one hand, we find puppet determinism, according to which the genes pull the strings and human beings dance. Puppet determinism is what Lewontin opposes. His opposition leads him simply to add a second puppeteer, namely, the social environment. On the other hand, the gene myth includes promethean determinism. As Prometheus stole fire from the gods, geneticists are stealing knowledge of the genome from nature. This knowledge will permit us to design and build a pair of technological scissors with which to cut the puppeteer's strings. Once we have cut the strings, we will retie them, and we the human race will become our own puppeteer. Will the genes then dance when we pull these strings?

The logical question is this: If the genes have been our puppeteer, did they themselves give us what we need now to cut the strings and take control? Is human freedom itself a gift of the genes? If not, then what is the condition that makes possible transcendence of our genetic inheritance? If we have been puppets all along, how can we suddenly become the puppeteer?

In the pages that follow we will look at the three legs of our chair: molecular biology with its Human Genome Project, behavioral genetics, and evolutionary psychology, also known as sociobiology. Each leg, as we will see, supports genetic determinism only in the face of demur. In each case belief in genetic determinism as human determinism is challenged as unsupportable scientifically. It is unsupportable scientifically for three reasons.

First, genetic determinism understood as human determinism relies upon a category mistake—that is, the mistake of confusing wholes and parts. Freedom is found at the level of the self or the human person as a whole organism; and a person as a whole is not reducible to the determinism of any or even all of his or her component parts such as genes. Genetic determinism, to the extent that it exists, is simply not in conflict with freedom experienced at the human level.

Second, at the molecular level genes are not determinative in any absolute or exhaustive sense. Genes interact in subtle but decisive ways with their environment. In addition, mutations and other similar changes in the DNA are not predictable but rather the result of chance.

Third, as our knowledge of genetic influence on human nature and well-being grows, we need to be open to the possibility of discovering that the genes themselves determine that we are free. A fascinating thought is this: it just may be the case that human freedom is itself the product of genetic determinism. At this point in history, therefore, we will have to conclude that widespread belief in the puppet determinism of human life is primarily a cultural phenomenon, what we are calling the gene myth.

In addition to sorting out the relationship between gene science and gene myth, we will subject both to the hermeneutic of secular experience and uncover the underlying ethical and theological dimensions. The controversy over the gene myth takes on such a sense of seriousness because it is being driven by an underlying ethical question: What should we do about human suffering? If we pursue genetic research toward the end of developing new medical technologies that could relieve human suffering, then will we in the process create new injustices? When we play God with DNA, do we risk creating new stigmas and, hence, more social suffering?

A theological analysis will show that this struggle is reflective of the fundamental tension that exists at the root of the human condition, namely, the tension between *soil* and *spirit*. This basic tension is rooted in finitude and sin, in the desire to transcend our biological roots and in the desire to transform evil into something good. We will look at the first of the three legs holding up the gene myth, molecular biology, and then explain this tension.

Molecular Biology

What is the relationship between belief in genetic determinism and the actual work of molecular biologists in the Human Genome Project seeking knowledge of DNA sequences and mapping the human genome? As we have said: genetic determinism enters the picture in the form of a conceptual lens or thought structure that frames the discussion regarding the significance of gene hunting and gene finding. A beautiful country landscape may be aesthetically pleasing when we are walking in it; but we think differently about it when it appears as a picture on a highway billboard framed by a cigarette advertisement. Determinism is a conceptual frame that we use to picture what is happening in genetic research.

A front page headline of *USA Today* provides a clear example: "Alzheimer's gene found." The article subtitle reads: "Mutations seen as cause of most aggressive cases."[4] The word "cause" here is significant. Genes *cause* disease, it is reported. This presumes a framework of puppet determinism in that what happens to our phenotype is thought to be dictated solely by our genotype. It also presumes a framework of promethean determinism, because the mood of the article conveys a sense of impending triumph, the sense that medical science will gain victory over this debilitating disease. Genetic engineering is the spear in the hand of the warrior who will slay the dragon of disease.

One form of the underlying mythical structure is that defective genes are akin to despotic dictators, wreaking physical havoc on their subjects and oppressing them in some cases with unspeakable suffering. The medical scientist is our knight in shining armor. The medical knight, like an intelligence agent, enters the castle of the despotic king through molecular research, learns the secrets of the DNA citadel, and then opens the gates so that the soldiers of therapy can enter and overcome.

Although the attention given to molecular biology has focused on progress in medical genetics, the logic of determinism has expanded the scope of popular thinking. Might our behavior be DNA-dictated? If some genes dictate diseases, perhaps other genes dictate human behavior. Eye color and hair color and other phenotypical traits are genetically caused, so how about tastes and dispositions and inclinations and desires and such? A January 2, 1996 *New York Times* article title reads: "Genetic Link Found to Personality Trait: Explanation for Impulsiveness." That complex and hard-to-control phenomenon of impulsiveness now has an explanation, a genetic explanation, says the newspaper article reporting on scientific research. What follows, of course, is the popular image of a massive intelligence operation seeking out the genes that govern obesity, alcoholism, aggressiveness, violence, and sexual orientation.

The molecular biologists working in the laboratory tend not to be as deterministic as the popular image. There are at least two reasons. First, the practicing scientists can distinguish methodological determinism from philosophical—that is, ontological—determinism. Methodological determinism simply frames research so that the scientist can dissect organisms into their component parts, look for the mechanisms that determine biological processes, and better understand targeted phenomena. Laboratory scientists look for the genetic determinants for specific biological traits. The reductionistic assumptions of research do not require them to advocate a universal philosophical determinism that claims everything is biologically determined. Molecular and cell biologist R. David Cole says that

> biologists do not believe in determinism by genes in the absolute sense. . . . The danger is that the non-scientific interpreter will be confused by the deterministic talk of most scientists, especially genome researchers. The conversation of most scientists is in fact couched in terms of determinism even though the majority of scientists probably do not believe that we are complete slaves to our genes and environmental history. The confusion arises from a failure to distinguish between philosophical determinism and the methodological or operational determinism of science.[5]

The second reason why molecular biologists are reluctant to embrace a thoroughgoing ontological determinism is that they observe a very important fact: genotype alone does not determine phenotype. Genes do not rule alone. Environment plays a significant role in gene expression. Environment for a geneticist includes the chemical life within the cell as well as the external world, which stimulates or fails to stimulate cell activity. The mechanisms for gene activation or repression are sensitive to chemical forces inside and outside the cell. DNA will behave differently in a person who is well fed than in one who is starving to death. Even though genes establish potentials and set limitations, genes alone cannot determine the actual outcome. So Cole can write, "The gene is not, however, a sufficient explanation. The genome predetermines all the potential states of being and behavior, but it does not predetermine the person to any one particular state."[6]

To make a human person, we need both genes and environment, nature and nurture. At minimum, nurture remains as important as nature. In opposing what he calls "genetic predestination," *Nature* editor John

Maddox draws us back to the basics: genotype underdetermines phenotype.

> The link between genotype and phenotype is not always unambiguous. A genotype may be a necessary, but not a sufficient, condition for the phenotype; the individual concerned inherits only a susceptibility for the phenotype ... it would be rash to deny that the missing ingredients may be aspects of nurture. ... Is it likely that some of the most labile aspects of the human phenotype, those constituting personality, will be exempt from similar and more subtle influences?[7]

This is the case at the individual level. At the species level, through natural selection, genomes adapt to the environment. The principle of adaptation shows that it's not all in the genes. To understand genetic change over time we need an interactive or ecological perspective. The persons we are today are the result of a long genetic and environmental history, a history that is still ongoing.

Such deliberation leads Cole to agree with Maddox and write against genetic predestination.

> There is no reason for the non-scientist to be intimidated by the success of the deterministic approach in elucidating the biological role of genes in human nature, and certainly no reason to be intimidated by any scientist who might try to convince us that determinism is all that is. Although the case for free will cannot be rigorously proven, those of us who believe in it need feel no threat from the findings of the Human Genome Initiative.[8]

Walter Gilbert makes the same point.

> Genetic information does not dictate everything about us. We are not slaves of that information. We must see beyond a first reaction that we are the consequences of our genes; that we are guilty of a crime because our genes made us do it; or that we are noble because our genes made us so. This shallow genetic determinism is unwise and untrue.[9]

Two observations are in order here. First, despite the fact that geneticists reflecting on their work at the molecular level tend not to be genetic determinists in the philosophical sense, the more popular interpretation of their science is. Determinism frames the cultural picture we have of genes, and the scientists themselves fit into this larger cultural picture.

Secondly, in repudiating genetic determinism, molecular biologists do not actually advance a full-fledged theory of human freedom. Rather, they simply buffer genetic determinism with environmental determinism. They end up with a two-part determinism, not actual freedom. Human freedom is simply not the subject matter they study. Molecular biologists study molecules. Freedom is a phenomenon that occurs at the level of the whole organism, at the level of the human self. To look only at the parts is to leave the whole out of the picture. This is not a crime, to be sure. It simply indicates that laboratory methods of genetic research are so focused on DNA that the subject matter that interests us here, human freedom, does not typically appear on their mental radarscopes.

It does appear on the mental radarscope of geneticist Francisco J. Ayala. He advances two arguments in behalf of the existence of free will, basing them both on common-sense evidence. "One is our personal experience, which indicates that the possibility to choose between alternatives is genuine rather than only apparent. The second consideration is that when we confront a given situation that requires action on our part, we are able mentally to explore alternative courses of action, thereby extending the field within which we can exercise our free will."[10]

These observations are of decisive importance. What we human beings experience every day is our freedom. We know when we have freedom, and we know when it is taken away. "Freedom" is a word we use to identify a certain form of experience we have as individuals and perhaps as societies. Philosophers in the past and present have sought to explain freedom with greater or lesser degrees of adequacy; but what they have sought to explain is something that already exists indelibly in human experience. When molecular biologists seek to track the influence of gene expression on human phenotype and on human behavior, they are contributing to an explanation of human freedom. However, their explaining it cannot destroy it. They cannot take it away. They cannot declare that it no longer exists. At best geneticists can offer a partial explanation of the factors that influence human behavior and, hence, human freedom. The existence of this freedom is simply a given that *inter alia* science tries to explain.

The puppet determinism in the gene myth risks confusing us because it risks committing the fallacy of misplaced concreteness. Alfred North Whitehead identified this fallacy as an accident in thought when we confuse what is concrete with what is abstract, when we confuse the fact in front of us with the theory to explain it.[11] The fact in front of us is that human beings experience freedom and know that they experience it. To deny this experienced reality on the basis of a reductionist method or on the basis of a hypothesis about genetic determinism is to grant the explanation priority over what we are trying to explain. To try to apply newly gained knowledge regarding genetic coding to human behavior and to reflect upon its implications for human freedom is well worth intellectual pursuit; but it would be a fallacy to begin with the idea that the freedom we experience does not exist on the basis of an attempt to explain it. Laboratory molecular biologists are less likely to commit the fallacy of misplaced concreteness than are the more popular advocates of the gene myth.

Behavioral Genetics

Behavioral genetics, much more than molecular biology, directly fosters the gene myth.[12] However, this need not be the case, as we shall see.

The task of behavioral geneticists is indicated by their name: they seek to identify and label as many behaviors determined by our genes as they can. Using statistics they study variations in patterns of human behavior and try to estimate their heritability. Granting that both nature and nurture are factors in human development, they assume that what cannot be accounted for genetically must be environmental.

Heritability is measured in percentages. It is assumed that eye color is determined totally by the genes. Therefore, the heritability of eye color is 100 percent. Height is due to genetic influence, to be sure; but nutrition also plays a role. Therefore, the heritability of height is 90 percent—that is, 90 percent of the variation of height is accounted for by corresponding variation in genomes, whereas 10 percent is accounted for by diet and other environmental factors.[13]

Studies of twins capture the imaginations of behavioral geneticists. For example, Thomas Bouchard's research at the University of Minnesota, wherein workers interviewed 50 pairs of identical twins (monozygotic or MZ twins) who had been separated shortly after birth and raised in separate environments, has stimulated considerable excitement.[14] On the question of IQ, previous studies had estimated the heritability of intelligence at roughly 50 percent. The University of Minnesota team raised that to 70 percent. Going beyond phenotypic traits to behavior, such studies claim to have identified underlying genetic influences on leisure-time interests, mate selection, proneness to divorce, job satisfaction, political orientation (conservative vs. liberal), and even religious preference.

Other studies report exciting discoveries of genetic power influencing our lives. Two British women, identical twins separated at birth and said to have been reunited only at the point of the study conducted by the behavioral geneticists, showed uncanny commonalities. Their first born sons were named Richard Andrew and Andrew Richard. In another case, two twin men, both named Jim, named their respective pet dogs "Toy," married women named Linda, divorced them and remarried women named Betty. Yet, another pair of twin men discovered they had both become firefighters and preferred Budweiser beer.[15] Behaviors as diverse as smoking habits, insomnia, marriage and divorce, careers, hobbies, use of contraceptives, coffee drinking, menstrual symptoms, and suicide have been found to have far higher rates of concordance for identical twins than for fraternal twins. This finding suggests that these traits are more influenced by genes than was previously suspected.[16]

Interpretations of twin studies in the popular culture range from a "wow, gee whiz" fascination to a more circumspect anxiety regarding freedom and the meaning of individual life. "The premise of free will is that we become the people we *choose* to be. Suppose, then, we meet an Other who is, in every outward respect, ourself." This is the dark cloud of nihilism Lawrence Wright sees floating above behavioral genetics. "Isn't there a sense of loss? A loss not only of identity but of purpose? We are left wondering not only who we are but why we are who we are."[17]

Lindon J. Eaves sees dark clouds, but not the nihilistic clouds that threaten to rain on human freedom. Rather, he sees cloudy vision regarding just what genetic science may or may not be leading us to. Eaves is Distinguished Professor of Human Genetics and Professor of Psychiatry at the Medical College of Virginia, Virginia Commonwealth University. He heads one of the largest twin studies research programs, known as the Virginia 30,000. Eaves believes in freedom, because he can see it. How did we human beings become free? Can we say the genes did it? Can we say the genes gave us freedom?

The common-sense assumption usually brought to this discussion is this: the genes make monozygotic twins identical, but it is freedom that makes them differ. If we find very little difference, then we might conclude that the genes are all-determinative. Eaves disagrees. Freedom may actually turn out to be the reason that monozygotic twins are alike; because twins differ for the most trivial reasons that have very little to do with our identity as human beings. Much more than just genes make MZ twins similar. He believes we need to investigate the relationship of freedom to all aspects of our identity as humans, including the biological *and* the cultural roots of variation.

It is important that we recognize that behavioral geneticists do not limit themselves to a single-factor determinism of genes alone. To speak of genetic effects is not to say that environment does not matter. If the environment were irrelevant, then the correlations for identical twins would be 100 percent. But certainly they are not. Typical variations run between 50 percent and 80 to 90 percent. These variations imply an environmental contribution. The point that should not be missed here is this: we should be surprised to find out that genetic factors play a role at all.

The road taken from genetic expression to human behavior is very long. The genes may affect our sensitivity to environmental changes and even influence our exposure to "good" or "bad" environments. The human mind—even the genetically influenced human mind—has a capacity to sift experience and create new environments. Adaptation is an interactive process of genes and environment. Behavioral geneticists are two-factor determinists.

The issue regarding the location of human freedom now becomes more subtle. To get at the subtlety, let us ask this question: Does human freedom transcend two-factor determinism—its biological and environmental origin—as a third factor? Or, is human freedom exhaustively explainable by appeal to genes and environment? And if it is exhaustively explainable, is it still freedom?

Philosopher Bruce R. Reichenbach and behavioral geneticist V. Elving Anderson state the issue regarding the prospect that human freedom transcends two-part determinism.

> The behavioral geneticist ascribes the cause of human behavior to a combination of genetic and environmental factors. What is not genetic or heritable derives from the environment. But what room then remains for human agency? . . . without choice there cannot be moral responsibility. But choice means that persons contribute something to the action over and above what derives from their genetic heritage and environmental input.[18]

Reichenbach and Anderson are suggesting that two-part determinism—genes plus environment—is inadequate for accounting for human free will. Human agency, it is assumed, must be a third factor, perhaps independent of the first two.

Eaves is a bit more wary of prematurely protecting a human freedom that seems to exist independently of its biological substrate. We may discover that genetic fingers reach right up into the very highest expressions of the human spirit. If such a biological determinism turns out to be demonstrable, what then?

Eaves seems to be able to have both maximum genetic determinism and human freedom. How can this be? Because the genes determine that we will be free, or at least determine that we humans at the level of the personal self will set out on a quest for freedom. For every strand of DNA that might constrain or inhibit us, there is another that makes us cry "freedom." Our natural endowment is by no means a prison, confining or limiting the human future. Rather, our biological inheritance has sown the seeds that will sprout as liberation. The human spirit is DNA's way of dreaming new futures. Eaves says, "I would argue that evolution has given us our freedom, that natural selection has placed in us the capacity to stand up and transcend the limitations of the environment. So, I think the quest for freedom is genetic. I can't prove it, but I think it's a way forward."[19]

Evolutionary Psychology (*alias* Sociobiology)

In the open market of saleable worldviews, one product of science-related theorizing is to draw countless new buyers by appealing simultaneously to our fascination with determinism as well as our thirst for liberal freedom. Some label it "evolutionary psychology." The prototype was first marketed as "sociobiology" with the publication of Edward O. Wilson's *Sociobiology* in 1975 and Richard Dawkins's *The Selfish Gene* in 1976. Perhaps because this earlier version was advertised too aggressively as a form of iconoclastic determinism, it did not capture much of the intellectual market. With design modifications and relabeling, sociobiology is back on sale under different names: behavioral ecology, Darwinian anthropology, evolutionary psychology, and evolutionary psychiatry. Its appeal is that it draws upon science but builds a worldview that is totalistic in its explanation of what is meaningful to our lives. "Here 'worldview' is meant quite literally," says science writer Robert Wright. "The new Darwinian synthesis is, like quantum physics or molecular biology, a body of scientific theory and fact; but, unlike them, it is also a way of seeing everyday life. Once truly grasped . . . it can entirely alter one's perception of social reality."[20] With evolutionary psychology we buy the gene myth in the large economy size. This warrants our looking at it a bit more thoroughly.

Wilson throws down the determinism gauntlet: "The agent itself is created by the interaction of the genes and the environment. It would appear that our freedom is only a self-delusion."[21] Significant here is not whether Wilson factors in only genes or genes plus environment. Rather, what is significant is that he assumes that determinism renders freedom a delusion. It may be a very useful delusion, but a delusion all the same. Eliminated from this calculus is the human person as a free agent who evaluates and makes decisions. Gone is the person who determines. Whether we operate with single-factor or double-factor determinism, the gene myth (inclusive of its detractors) systmatically factors out the human self.

This is what gives evolutionary psychology its universal scope. By emphasizing human unity and continuity across cultures, evolutionary psychology marks a counterpoint to ideological pluralism characteristic of some forms of postmodernist thought. "Today's Darwinian anthropologists, in scanning the world's peoples, focus less on surface differences among cultures than on deep unities. Beneath the global crazy quilt of rituals and customs, they see recurring patterns in the structure of family, friendship, politics, courtship, morality. They believe the evolutionary design of human beings explains these patterns." This is a view of the human race in its universal scope, an anthropology based on nature and the scientific perception of nature. Why is it that all peoples have a deep sense of justice—justice understood as "one good turn deserves another" or "an eye for an eye, a tooth for a tooth"—that is, the sense of retributive justice that shapes human life everywhere on this planet? The answer of evolutionary psychology: the evolution of our DNA has brought the entire human race to this point.[22]

According to this theory, the evolutionary driver has been and continues to be the selfish gene. What this means simply is that genes seek unashamedly one destiny: namely, to replicate themselves and to go on re-replicating themselves forever. So, they drive the organism within which they reside to reproduction, to maximum fertility. "The organism is only DNA's way of making more DNA," says Wilson.[23]

Why are our minds preoccupied with sex? Evolutionary psychologists have the answer: survival of the genes. Why does planet earth have an overpopulation problem? Same answer. Why do clans and races and nations go to war and commit genocide to eliminate other people who differ genetically? Same answer. Why do some people among us feel altruistically about helping others to whom they are not genetically related? Same answer.

Genes produce memes, adds Dawkins. Memes are cultural products and analogues to genes. Genes are biological, whereas memes are cultural. An idea-meme is a mental entity capable of being transmitted from one brain to another, one generation to another. Memes travel via social inheritance. Memes, like genes, are selfish. They try to dominate the brain's thought processes, and they do so at the expense of rival memes. Memes exploit the culture to their own survival advantage. This gene-meme combination certainly echoes puppet determinism. Yet, Dawkins lives in a liberal society and evidently feels he must give voice to liberal views about human freedom. So he adds a version of the understanding-decision-control formula: once we, through science, understand how genes determine us, we will then gain the power to steer our human future. "We have the power to defy the selfish genes of our birth and, if necessary, the selfish memes of our indoctrination."[24] Without even a blink, Dawkins jumps from puppet to promethean determinism. So does Wilson. He asserts that, even though our genes may be both selfish and determinative, we can be altruistic. Human nature can embrace more encompassing forms of altruism and social justice. Genetic biases can be trespassed and ethics altered.

Are We Completely Determined by the Selfish Gene?

One of the frequently discussed problems in sociobiological theory is this: Once it is established that the genes are in charge and the genes are selfish, how can we explain nonselfish or altruistic human behavior? Or, even more difficult: How can we *describe* ourselves as driven by selfish genes and then *prescribe* nonselfish or altruistic morality? As fascinating as this issue is, it is not the one I wish to address at this point. Rather, I want to pursue the tenacious adherence on the part of sociobiologists to genetic determinism as an explanation for all human behavior.

The excitement elicited by the sociobiological philosophy is due to its attempt to explain culture—which we experience as an expression of human freedom—by appeal to biological processes, which are presumed to be deterministic. "The genes hold culture on a leash," writes Wilson. "The leash is very long but inevitably values will be constrained in accordance with their effects on the human gene pool."[25] He goes on to argue that the human brain is a product of evolution. Human behavior, including our deep capacities for emotion, is the circuitous technique by which our genetic material has been and will be kept intact. Although culture provides us with an apparent array of options from which we can make choices, our genes constrain and circumscribe and overrule our behavior. The central tenet of sociobiology or evolutionary psychology is that our behaviors are shaped by natural selection in our genetic history.[26] This applies to all aspects of culture including ethics and religion—even the altruistic ideals of religion. The sociobiologist does not find ethical norms already present in biological evolution, but he or she argues that genetic evolution predisposes us to accept certain moral norms—namely, those consistent with the objectives of natural selection. The tenacious adherence to sociobiological theory is that this naturalistic theory appears to have the power to explain the broad scope of human behavior, even inconsistent patterns of behavior such as selfishness and altruism.

It is instructive to look at the specific arguments raised by evolutionary psychologists that connect the selfish gene with morality at the cultural level. Wilson and Michael Ruse state the theory starkly: "Morality, or more strictly, our belief in morality, is merely an adaptation put in place to further our reproductive ends. Hence the basis of ethics does not lie in God's will . . . or any other part of the framework of the universe. In an important sense, ethics . . . is an illusion fobbed off on us by our genes to get us to cooperate."[27] That is, ethics makes us fertile, and fertility is the means whereby genes and memes survive. We human beings may think we are in the driver's seat steering culture with a moral steering wheel, but in fact we are puppets, and the genes are pulling the strings.

The arguments of Wright are indicative of this view. They are also important for the gene myth. Wright's views have made it big in the media. A *Time* cover draws attention to his article by picturing a broken wedding ring with the words in large letters, "Infidelity: It may be in our genes."[28] His article in *The New Yorker* is titled, "The Biology of Violence."[29] Appeals to DNA explain our penchant for immorality. Appeals to DNA also underwrite our desire for aspiring to live the moral life. Are these really compatible?

Evolutionary psychologists do not just describe, they prescribe. They pursue ethics. Evolutionary ethics is founded upon, or at least informed by, genetic determinism. It is axiomatic to the likes of Wright that human morality is itself a product of evolutionary history and that ethics has developed to further the evolutionary process. "People tend to pass the sorts of moral judgments that help move their genes into the next generation."[30] Or, "what is in our genes' interests is what seems right—morally right, objectively right."[31] This is an overtly nontheological naturalism. Here, morality is based in nature, strictly nature, not God. "There's definitely no reason to assume that existing moral codes reflect some higher truth apprehended via divine inspiration or detached philosophical inquiry."[32]

What about the naturalistic fallacy? Can we draw an *ought* from what *is?* The tacit agreement among ethicists is no, *is* is insufficient to ground *ought.*

What is already does not ground what ought to be and is not. A moral sense of what ought to be is a *prescription* that adds something beyond a *description.* Yet, it would appear that when evolutionary psychology moves from description to prescription, from science to ethics, from nature to morality, that it is committing just this fallacy.[33] Wright acknowledges the risk of committing this fallacy when making such a move; and he agrees that nature itself cannot be a moral authority. If in nature it appears that might makes right, for example, we human beings need not follow this principle. "To say that something is natural is not to say that it is good. There is no reason to adopt natural selection's values as our own."[34] Nevertheless, he argues that morality needs to be based on a true understanding of human nature and, of course, evolutionary psychology provides that understanding.[35]

> Altruism, compassion, empathy, love, conscience, the sense of justice—all of these things, the things that hold society together, the things that allow our species to think so highly of itself, can now confidently be said to have a firm genetic basis. That's the good news. The bad news is that, although these things are in some ways blessings for humanity as a whole, they didn't evolve for the good of the species and aren't reliably employed to that end.[36]

The process of natural selection is blind to human values, even though evolution appears to be designing things for a purpose.[37] The doctrine of natural selection holds that if within a species variation exists among individuals in their hereditary traits, and if some traits are more conducive to survival and reproduction than others, then those traits will become more widespread in the population. Darwin summed up natural selection in ten words, "multiply, vary, let the strongest live and the weakest die."[38] Although Darwin did not himself coin the phrase, "survival of the fittest," he accepted it; and today evolutionary psychologists use the term "fittest" to refer to adaptation of genes to environment. More specifically, the DNA of our ancestors adapted to the EEA, the "environment of evolutionary adaptation" or the "ancestral environment." This leads to what Wright identifies as the cardinal principle of sociobiology: "[T]hose genes that are conducive to the survival and reproduction *of copies of themselves* are the genes that win."[39] Wright continues:

> Natural selection "wants" us to behave in certain ways. But, so long as we comply, it doesn't care whether we are made happy or sad in the process, whether we get physically mangled, even whether we die. The only thing natural selection ultimately "wants" to keep in good shape is the information in our genes, and it will countenance any suffering on our part that serves this

purpose.[40]

It is this adaptation process that leads to the illusion that natural selection is consciously designing organisms. This spawns a key move in the method of evolutionary psychology, namely, "pretending you're in charge of organism design" in order to figure out "which tendencies evolution is likely to have ingrained in people and other animals."[41]

This is where the determinism-freedom axis comes into play. It invokes the unquestioned assumption of our liberal culture that knowledge of what determines us gives us the power to make us free. Here is how it works. Take for example our sexual proclivities such as attraction to the opposite sex. This is driven, according to sociobiological theory, by the unconscious working in us of natural selection. Unconsciously, what we experience as flirtation and love is in fact being driven by the evolutionary force of adaptation through reproduction. "Lust and other such feelings are natural selection's way of getting us to act as if we wanted lots of offspring and knew how to get them, whether or not we actually do."[42] Or, "the reason they want children is because their genes want children."[43] Or, "emotions are just evolution's executioners."[44] In short, our sex lives are determined by the demand of our DNA to replicate itself through reproduction. Wright offers more thought on this:

> Everyday experience suggests that natural selection has wielded its influence largely via the emotional spigots that turn on and off such feelings as tentative attraction, fierce passion, and swoon-inducing infatuation. A woman doesn't typically size up a man and think: "He seems like a worthy contributor to my genetic legacy." She just sizes him up and feels attracted to him—or doesn't. All the "thinking" has been done—unconsciously, metaphorically—by natural selection. Genes leading to attractions that wound up being good for her ancestors' genetic legacies have flourished, and those leading to less productive attractions have not.[45]

Ex post facto, Wright can argue that our sexual attractions today are the product of our ancestors' adaption to the environment for the purposes of maximizing reproduction. This is alleged to be true whether we realize it or not. In order for us to understand what is really going on within sexual attraction, we need sociobiological theory.

Not only are our sexual proclivities unconsciously driven by the need of the DNA to replicate itself, all cultural values including moral values are similarly determined by what the genes want. Take happiness, for example, an item that appears quite high on ethical scales of value. Wright says that "the basic mechanism by which our genes control us is the deep, often unspoken (even unthought), conviction that our happiness is special. We are designed not to worry about anyone else's happiness, except in the sort of cases where such worrying has, during evolution, benefited our genes."[46] The pursuit of one's own happiness, like the other items we value, is adaptive; and we are determined to pursue happiness because it—the pursuit, not happiness per se—is adaptive.

When we raise the adaptive process from our unconscious to our conscious awareness, two things happen. First, we become aware that we are determined. We are puppets, and our genes are the puppeteers. Second, we become liberated from the power of the puppeteers.

> Understanding the often unconscious nature of genetic control is the first step toward understanding that—in many realms, not just sex—we're all puppets, and our best hope for even partial liberation is to try to decipher the logic of the puppeteer.[47]

This method I have earlier identified as the understanding-decision-control formula, whereby the nearly invisible Enlightenment assumptions regarding natural determinism and human power come to paradoxical expression. To the modern Enlightenment mind, determinism in nature—whether genetic or in other forms—poses no threat. We are fearless in the face of this form of determinism. Why? Because we believe that through science, we can understand how determinism works; and through technology, we can steal that very determining power and press it into the service of distinctively human needs and desires. Understanding nature and making decisions regarding how we should employ nature's processes gives us control over our destinies, so we think. What Wright has done here is to take the first step: he employs his own scientifically based theory to understand how the genetic puppeteer works, and then immediately begins to talk about liberation.

The vocabulary tumbles all over itself in such discussions, but it is clear that determinism—whether one-part or two-part determinism—remains the mainstay. In objecting to genetic determinism alone, Wright argues for genes plus environment. The phrase "genetic determinism" exudes "ignorance as to what the new Darwinism is about . . . everyone (including Darwin) is a victim not of genes, but of genes and environment together: knobs and tunings."[48] It looks like we have a return to nature plus nurture. But, because all environmental influences are mediated by our biology, Wright goes on to clarify: "even though the term *genetic determinism* is confused, the term *biological determinism* isn't."[49] Wright can describe the environment as biological, because environmental adaptation through natural selection evolutionarily alters the genome. Make no mistake about the conclusion. This is a deterministic scheme, whether it be a one-factor or two-factor determinism. The reason we feel that we are free at the conscious level is that we are ignorant of all the determining factors at our unconscious biological level.

Even more to the point: our lack of conscious awareness regarding what determines us is no accident. It is part of nature's design. We have been programed by our genes to be ignorant of their power over us. The delusion of free will with which we operate is a delusion perpetrated on our consciousness because it serves the designs of genetic interest, namely, adaptation.

Wright himself believes that free will is a "useful fiction." As a utilitarian, he finds the fictional belief in free will useful for limiting crime. If we presume that each person is free and therefore responsible for his or her behavior, and if we prescribe certain punishments for certain antisocial behaviors, potential criminals may be deterred. To make the legal system work, we must ascribe responsibility in order to apply punishment, and it is the threat of punishment that will reduce the number of crimes committed. The net good for the greatest number in our society is enhanced.

As a corollary, Wright does not want biological determinism to become a form of excuse that will exculpate criminals from their responsibility. He has no sympathy for defense appeals to diminished capacity, temporary insanity, post-traumatic stress disorder, and such biological arguments that mess up the crime-punishment relationship. We cannot claim innocence on the grounds that "my genes made me do it." To tell people that they are not to blame for their antisocial behavior is only to make future misbehavior more likely. If we are to have civility and justice in society, we must sustain cultural support for the belief that we are responsible for what we do. In other words, it is vital that we continue to believe the fiction of free will if we are to pursue the greater good.

Can our genes predispose us to evil? Can we speak of a genetic version of original sin? Yes, says Wright. Even though the concept of evil does not fit squarely into our modern scientific point of view, he acknowledges, believing in evil is still a useful fiction. It is useful because our genes predispose us toward behaviors that fail to lead toward our happiness. Note that Wright is a utilitarian—a utilitarian is a hedonist who advocates the greatest happiness for the greatest number—of the J. S. Mill type for whom happiness is the highest good.

> There is indeed a force devoted to enticing us into various plea-sures that are (or once were) in our genetic interests but do not bring long-term happiness to us and may bring great suffering to others. You could call that force the ghost of natural selection. More concretely, you could call it our genes (some of our genes, at least). If it will help to actually use the word evil, there's no reason not to.[50]

Yet (and this is the paradox) there seems to be no fear here that bio-logical determinism eliminates human freedom. Again Wright offers:

> If love of children is just defense of our DNA, if helping a friend is just payment for services rendered, if compassion for the down-trodden is just bargain-hunting—then what is there to be proud of? One answer is: Darwin-like behavior. Go above and beyond the call of a smoothly functioning conscience; help those who aren't likely to help you in return, and do so when nobody's watching. This is one way to be a truly moral animal.[51]

In other words, although we are determined by our DNA, we are still free to rise above our DNA, and the way we rise above is by becoming moral. And what moral means here is clear: engaging in acts of love for which no personal gain—or reproductive gain for one's genes—is the goal.

The reader might be saying now that this seems like such a contradic-tion. If our DNA is determinative, and if it determines our behavior at the unconscious level, and if human morality is but a fiction employed by the DNA to attain its own reproductive ends, then how can we use morality to liberate ourselves from DNA determination? How can the puppet liberate itself from the strings of the puppeteer? Certainly Wright's Darwin-like view of behavior, if the theory holds any water, is just as much a product of unconscious DNA determination as kin altruism. That this is a contradiction is blatant. Yet, I can only report that this particular contradiction appears again and again in the case made by sociobiolo-gists. It appears that evolutionary psychologists, once they have made the deterministic argument, suddenly wake up and realize that they live in a liberal society; and then they want to speak of liberty. Freedom gets reaffirmed, and liberation becomes our task. It's like inviting prisoners in a concentration camp to a freedom pep rally, and then when it's over, asking them to return to their cells.

Sociobiology's Critics

Criticisms of sociobiology or evolutionary psychology are not hard to find. Among the more vitriolic critics we find Richard Lewontin and his colleagues. Sociobiology has popular appeal, they argue, because it legitimates the status quo. Sociobiologists commit the naturalistic fallacy, with the effect that *is* abolishes *ought*. This means that a social philosophy based on genetic determinism nullifies our sense of guilt and responsibility. "It is precisely because biological determinism is exculpatory that it has such wide appeal," they write. "If men dominate women, it is because they must. If employers exploit their workers, it is because evolution has built into us the genes for entrepreneurial activity. If we kill each other in war, it is the force of our genes for territoriality, xenophobia, tribalism, and aggression. Such a theory can become a powerful weapon in the hands of idealogues who protect an embattled social organization by a genetic defense of the free market."[52] Evolutionary psychologists such as Michael Ruse or Robert Wright would object to such criticisms, denying that they commit the naturalistic fallacy. And certainly Edward O. Wilson and Richard Dawkins embrace Lewontin's liberal values. Yet, the Lewontin critique remains on target, because the fundamental premise of genetic determinism, especially in its puppet form, leads to laissez-faire fatalism.

Intellectual war correspondent Philip Kitcher clarifies the battle lines between the friends and foes of sociobiology. In the smoke of battle we may mistakenly think the war is being fought over single-factor or double-factor determinism—that is, over genes alone or genes plus cultural environment. Wilson's enemies such as Lewontin and others ascribe to Wilson a view of "the iron hand of the genes," but this does not apply. Although Wilson—along with contemporary evolutionary psychologists such as Wright—emphasize genetic determinism, they readily acknowledge that evolutionary history is characterized by genes and culture in mutual interaction. When the sociobiologists retaliate, they ascribe to their environmentalist opponents the view that we human beings are born with a blank mind or *tabula rasa,* that we are infinitely malleable. But this does not apply either. Both armies are contending for the spoils of genes-environment determinism. "What initially appears as a furious debate quickly dissolves into a tempest in a teapot," writes Kitcher. "Pop sociobiologists and their opponents agree that genes and environment together determine phenotype, and that is the end of the matter."[53] Yet there is still a stand off, even if it is not simply between bald genetic determinism and bald cultural determinism. The genotype is fixed. The phenotype is not. Within a limited array of possibilities, the genotype will assign the phenotype different functions in different environments. The sociobiologists believe that these phenotypic functions vary relatively little, and that when they do, it is only in response to a drastic change in environment. For the sociobiologist the phenotypic strings may be a bit loose, but the genetic puppeteer still pulls them. The critics, in contrast, maintain that the functions are quite responsive to changes in environmental variables. Culture contributes, perhaps even controls. The dispute is over relative emphasis. In sum, although both armies want to capture the nature-nurture turf, sociobiologists still want to plant the genetic flag while their opponents the environmental flag.

What is of primary concern for us here is another point that Kitcher makes. "Freedom does not consist in the absence of determination but in the way in which the action is determined," he says.[54] To posit randomness or indeterminism at the biological level would not bring us one step closer to understanding human freedom. This is because freedom must be seen as a form of determination—that is, determination by a human subject. To act freely is to make choices in light of evaluation. To act freely is to determine an action based on what one values, and to do so without enslavement to compulsions from within or constraints from without.

To evaluate options during the process of making choices presupposes value and purpose. It is fair to ask whether or not our human experience of value or purpose belongs only at the human level of the self, or whether values and purposes are built into our genetic code. A widespread tacit agreement seems to be that evaluation belongs strictly to the human level where culture plays a role, not at the prehuman biological level. Philosopher Holmes Rolston III and theologian Langdon Gilkey address this issue.

Among the more sympathetic critics of sociobiology we find Holmes Rolston III. Rolston is critical of the selfish-gene theory, because he does not like the moral category of selfishness applied to premoral nature. Nature is amoral, but this is not to disparage it. Genes in DNA do in fact perpetuate themselves through multiple generations as Dawkins and others describe. Rolston prefers to describe this as biological conservation. For an agent to be selfish it must have a self. Genes alone, considered apart from the whole organism, simply are not selves. Because genes are not moral agents as selves are, they cannot be either selfish or altruistic.[55]

Rolston makes one of the chief points of our study here, namely, that moral agency occurs only at the level of the self; and what constitutes a whole self cannot be reduced to its genetic parts. Yet, his objection to the phrase "selfish gene" may be excessive. Evolutionary psychologists describe the genes as selfish by analogy to our experience of human self-ishness, but the analogy is readily accepted as partial rather than total. In defense of the evolutionary psychologists, the use of the language of self-ishness at the DNA level belongs to the larger argument that human cul-ture, including its morality, is rooted in our biological makeup. It is not an attempt to superimpose a human outrage at genetic immorality, but rather an attempt to show that moral agency at the cultural level is a predictable expression of a more primitive form of agency at the molecular level.

Rolston also believes that culture transcends its bioevolutionary his-tory. He believes in an exodus of spirit, wherein we begin as matter and energy and then, through a long process of biocultural evolution, emerge with the spirit of freedom. Freedom belongs to human belief. Genes do not finally determine what we believe. Rather, they produce brains and minds that can be taught anything. Cultural selection, not a DNA pup-peteer, determines our beliefs. The beliefs that persist through many gen-erations are the culturally adaptive ones. This ought not to be called Darwinian selection. Rather, it is Lamarckian selection. Better still, we should refer to it as cultural selection, recognizing that it transcends our biology. Culture is launched, but certainly not leashed, by our DNA.

A parallel criticism of the work of sociobiologists is Langdon Gilkey's exposure of equivocation in their analogical use of language. Gilkey observes, as does Rolston, that the evolutionary theorists borrow lan-guage from the realm where human freedom is found and then, by anal-ogy, apply it to the mechanistic interpretation of natural processes. Then, they conclude, human freedom is governed by more primitive mechanis-tic processes and, hence, does not exist (at least as anything more than an illusion). This is fallacious, of course.

Gilkey says sociobiologists use analogy at several undiscriminated levels. At level one, they borrow the universal language of purpose and apply it to biological processes with phrases such as "in order to" or "such and such is developed for this or that function." But a cardinal doc-trine of evolutionary theory is that nature has no purpose, no built in tele-ology. So, we have a nonpurposive application of purposive language. Second, sociobiologists make specific analogy to *conscious* purposes in human existence to articulate *unconscious* yet apparently purposive behavior in DNA. Third, the analogy marks a shift from the conscious and unconscious purposes of the organism as a whole, as a phenotype, to the unconscious purposes of a part of the organism, the genes in the geno-type. The gene is not the organism, yet qualities such as drive or purpose ascribed to the organism as a whole are applied to one element within the organism, namely, the gene.

What we end up with is a nonpurposive application of purposive lan-guage, a mechanistic application of language to DNA that ordinarily operates with reference to the whole human organism. In the process of crossing the bridge from human experience to biological description, something is left behind, namely, freedom. No wonder sociobiologists do not find freedom on the other side. So, Gilkey asks rhetorically, "Where is the analog, if there is one to that sense of freedom, felt from the inside, when we choose our purposes and feel our responsibilities? If there is none, perhaps the analogy fails."[56] The linguistic confusion in the field of evolutionary psychology is immense.

Despite all the smoke and mirrors about altruism, evolutionary psy-chology cannot locate solid ground upon which to construct a compelling ethic. Evangelical critic Stanton L. Jones argues that Robert Wright, like the utilitarianism to which he ascribes, cannot "articulate any persuasive reason why, when my interests run counter to the general good, I should sacrifice my advancement for the welfare of others. The mindset of evo-lutionary psychology merely compounds this problem." Then, appealing to the divinely provided ground for altruistic love and service, Jones writes, "perhaps there is a basis for parents to love an adopted child, for a husband to remain faithful to his wife, or even for someone to forgo marriage and serve God as a celibate single. Only a view of persons as cre-ated beings can make sense of these human drives."[57]

Finally, we need to ask if a sociobiological vision of human nature is inimical to theological understanding or if these could be made compati-ble. Arthur Peacocke believes theologians could learn from the sociobiol-ogists, and the latter could learn from the former. Peacocke argues that theologians would do well to recognize that human nature at its founda-tion is biological and genetic, however much it is overlaid by nurture and culture. And further, by incorporating an evolutionary view we should acknowledge that the human being we see today has actually been created by God through a long evolutionary process. Through evolution God has presented us with freedom and its accompanying potentials for fulfill-ment. "God has made human beings thus with their genetically con-strained behaviour—but, through the freedom God has allowed to evolve in such creatures, he has also opened up new possibilities of self-fulfill-ment, creativity, and openness to the future that requires a language other than that of genetics to elaborate and express."[58]

Peacocke's objection to sociobiology is to its reductionism. There is no need for the methodological reductionism necessary for biological research to be translated into a naturalistic determinism that eliminates God. A theological understanding of evolution is called for in our time, and both sociobiologists and theologians should listen. As Peacocke says,

> Where the Christian theist differs from the sociobiologist, as such, is in his affirmation of God as "primary cause" or ground of being of the whole evolutionary process and, indeed, of God as the agent in, with, and under this process of creation through time. What constitutes the challenge to theology is a new apprehension and explication of God's presence and agency in the processes that biology, in general, and sociobiology, in particular have unveiled.[59]

My task here is not to grant thumbs up or thumbs down to the field of evolutionary psychology. It is certainly a daring and comprehensive form of speculative genetics, and it must be admired for its protean ideas. However, it ought not to be confused with natural science. It is a speculative philosophy striving for ideas of universal applicability. It is not scientific because it prescribes no experiments which could confirm or falsify its claims.[60] When confronted with blatant contradictions in human behavior, it presses on unabashedly, explaining them all by the same principles of genetic selfishness or natural selection. It appears to me to be a collection of *ad hoc* and *post hoc* arguments that are certainly fascinating in imaginativeness. Rather than science, however, this field is a filter through which science is culturally interpreted. Sociobiology, in its present generation as evolutionary psychology, is itself the gene myth at work in one of its grandest forms.

Reductionism in the Gene Myth

Confusion fertilizes. It is the confusion in the way language uses us that fertilizes the growth of the gene myth. The central problem of the gene myth is the problem of perception and reality, of public perception versus laboratory truth. Regardless of what scientists discover at their lab benches, culturally the gene appears to be all-determining. As we mentioned earlier, this deterministic conception has been labeled by some the "Strong Genetic Principle," by others "genetic essentialism," and by still others the "gene myth."

The gene myth emerges from two sources, the scientists themselves and the press reporting on the science. The scientists may have overstated the case. Ethicist Evelyne Shuster objects to this. She says that the leaders of the Human Genome Project, especially in the early years, described their work in deterministic and reductionist language. "The conceit is that once the structure and function of the genome is understood, once the concepts of genetic code, program, and messages are grasped, it may seem possible to have a gene-based explanation of all phenotypic characteristics, including all aspects of human health, disease, and even behavior."[61]

However, as we have seen, it is generally acknowledged among molecular biologists that genotype does not exhaustively determine phenotype. Nevertheless, a deterministic view leaves the lab and enters public consciousness, eliciting fear. To think that we human beings can be fully accounted for solely from our molecular structure "has been the central fear, and a reason for society's mistrust of the new biology. This is because reducing humans to molecular components, and the body (once cultural) to biochemical reactions, changes the way we think about ourselves as unique individuals, lessens the value of life, and undermines the notions of individual worth, freedom, and responsibility."[62]

That some of the molecular scientists are themselves tempted to believe the gene myth is due to a failure to distinguish between methodological reductionism and ontological reductionism. We simply must grant that reductionism as a method that focuses on the genes as the object of research has been extremely valuable in the production of new knowledge. No one would want to surrender it. Yet, to extrapolate on the success of what is being learned about genes in order to build a philosophical worldview—a worldview that says if we only keep studying DNA we will find that "it's all in the genes"—commits the fallacy of hasty generalization. Human being—that is, human ontology—may very well be the result of numerous factors, only one of which is DNA.[63] To know for sure, we need to study humanity at all levels from the cell on up through physiology to society and even to spirit.[64]

Ontological reductionism also misleads interpreters of the science, such as the press. The press looks for headlines that emphasize breakthroughs and revolutions just around the corner, and this attracts headlines that tell us "it's all in the genes!"

The method for identifying the gene myth is to examine those headlines. That is, to find the gene myth we need to examine culture. We do not look strictly at what goes on at the lab bench. By "culture" here I am referring to the mytho-memes or conceptual sets that float through the media and form our thinking about issues. Sources of these conceptual sets include cliches, newspaper and magazine headlines announcing genetic discoveries, issues in political debate, legal decisions, radio talk show con-

tent, soap opera plots, advertising images, musical lyrics, cartoon subject
matter, and oral jokes. In the media DNA—visually caricatured as a
twisted ladder in the print media and as a hundred-foot-high multicolored
double helix set in a fountain pond at Disney's Epcot Center in Florida—
is said to contain selfish genes, intelligence genes, genes that make us fat,
genes that make us aggressive, the "criminal chromosome," the "gay
gene," schizophrenia genes, alcoholism genes, and pleasure-seeking genes.
Determining not just our bodies but also our moral behavior, DNA apparently provides the genes that define our identity. Incapable of deceiving us,
DNA seems to have become the locus of what we used to think was our
true self.

"These popular images convey a striking picture of the gene as powerful, deterministic, and central to an understanding of both everyday
behavior and the secret of life," write Dorothy Nelkin and M. Susan
Lindee in *The DNA Mystique*. "Genetic essentialism," they go on,
"reduces the self to a molecular entity, equating human beings, in all their
social, historical, and moral complexity, with their genes."[65] In sum, to
find the gene myth we look at culture, and what we find in culture is a
symbolic gene that determines who we are and what we can do. Nelkin
and Lindee continue:

> Clearly the gene of popular culture is not a biological entity.
> Though it *refers* to a biological construct and derives its cultural
> power from science, its symbolic meaning is independent of bio-
> logical definitions. The gene is, rather, a symbol, a metaphor, a
> convenient way to define personhood, identity, and relationships
> in socially meaningful ways.[66]

Nelkin and Lindee make the case that DNA and its genes have garnered a sense of sacredness in our culture. Three qualities nominate our
genome for sacredness: its soul-like quality, its potential for immortality,
and its belonging to God's domain. First, "DNA has assumed a cultural
meaning similar to that of the Biblical soul . . . DNA, like the soul, bears
the marks of good and evil: A man may look fine to the outside world, but
despite appearances, if he is evil, it will be marked in his soul—or his
genes." Secondly, in addition to the essential truth of who we are behind
mere appearances, DNA as soul may indicate a brand of immortality.
They refer us to Michael Crichton's science fiction novel of movie fame,
Jurassic Park, wherein long extinct dinosaurs are brought to life again.
"DNA also appears to be immortal, containing within it everything
needed to bring the body back."[67]

Thirdly, Nelkin and Lindee engage in a brief analysis of "playing God." They identify our fear that lurking behind every genetic dream come true is a *Brave New World* type of nightmare. They note how in films such as *Jurassic Park* that the Frankenstein myth returns. Movies depict what horrible consequences we can expect when through the *hubris* of science gone amuck, we transgress our limits and enter forbidden territory. Like divine retribution, nature strikes back with merciless fury. Gene manipulation is a violation of the sanctity of life. Nelkin and Lindee note how articles in religious magazines and how public protests against genetic engineering express concerns about "tinkering" or "tampering" with genes. "To manipulate genes is to move them to the profane realm of engineering and technology. This, it is feared, will compromise their spiritual status. By opposing genetic engineering in these terms, such statements acknowledge the sacred power of DNA."[68]

Nelkin and Lindee almost, but not quite, articulate the incoherent double character of genetic determinism in the gene myth. On the one hand, puppet determinism is reflected in their observation: "Traits that are genetic appear as immutable, deeply resistant to change initiated through individual action or external intervention. . . . The ideology of genetic essentialism encourages submission to nature and to constraints on the possibilities for social change."[69] On the other hand, they implicitly recognize the presence of promethean determinism, giving it their own label, "genetic futurism."

Under the label "genetic futurism," Nelkin and Lindee identify four related principles that guide the cultural version of promethean determinism. The first is "the poor breed more" aversion. It begins with the widespread belief that poor people make more babies than rich people. When we mix in a little genetic determinism and make the assumption that poor people are that way at least in part due to inferior genomes, thoughts of population control begin to surface. CBS TV commentator Andy Rooney made a troublesome remark in 1990 that brought him a temporary suspension: "Most people are born with equal intelligence but blacks have watered down their genes because the less intelligent ones are the ones that have the most children."[70] A version of the concept of watered-down genes drove the eugenics movement in England, America, and Nazi Germany. The eugenics philosophy was to increase the proportion of children born to intelligent and healthy parents and to reduce the number of children born to feebleminded or unhealthy parents and to eliminate the inherited propensity toward poverty and criminality. Eugenics then and now takes the science of inheritance and the technology of reproduction and quickly subjects them to the interests of racism, classism, and xenophobia.

The second principle of genetic futurism is the elimination of "lives not worth living." With the growing ability to test the genomes of the unborn still in the womb to find genetic predispositions to disease, more and more fetuses will be selectively aborted. Those with disease-causing genes, especially if the disease in question would be an expensive one that medical insurance companies wish to avoid, will be aborted. People with disabilities will be prevented from entering the world. Those with healthy looking genomes will be brought to term. We can forecast an emerging need to set a criterion, perhaps a base line of potential health that an unborn child must meet before gaining permission to be born. "When disabilities are understood as economic burdens for the larger society, whether or not to bear a potentially imperfect child becomes a social as well as individual choice."[71]

The third principle supporting conscious control of our genomic future is racial defense against "the threat of extinction." This principle presupposes the theory of evolution with its concept of the survival of the fittest. Nature culls out the weak, the meek, the misfits, and the degenerate; and we should cooperate with nature to be sure that survival belongs to the strongest, the healthiest, and the most competent. At least this is the way neo-Nazi and similar right-wing racist groups such as White Aryan Resistance and the Skinheads see it. Governing reproduction so that the preferred race survives has become a political objective. On this point the racial purity groups share something in common with many ecology groups. While the racial purists want to select which race survives, the ecologists want to limit the total number of natural human survivors in order to avert eco-catastrophe due to overpopulation. What both share is the plan to govern reproduction to enhance the quality of human life on the planet. One ecology button reads: "Gene Police: You out of the pool."[72]

The fourth principle follows from the first three, namely, "limiting reproductive rights." Distressing to Nelkin is the fact that *The Bell Curve* by Richard Herrnstein and Charles Murray made it to the best-seller list in 1994. The thesis that high individual IQ correlates positively with social and economic success in American society coupled with the thesis that IQ is genetically determined could only lead in one direction for public policy: we need to allocate more resources for the wealthy and successful in our society and withdraw support from the lower classes. Lower-class people have lower IQs and, therefore, should be discouraged from having more children. Withdrawal of government-supported welfare and educational programs can now be scientifically justified. And, to take the argument further, mothers who would have received welfare should now receive Norplant, a medically inserted birth-control device. This precedent would place us on the road toward regulated reproductive rights, and any coherent reproductive policy that is informed by genetics would inevitably become eugenics.

The good news, at least according to Nelkin and Lindee, is that legislative proposals to govern reproductive rights through use of Norplant have gone down to defeat. This means that genetic futurism is not yet a government ideology. Yet it is gaining purchase in our culture. "Commonly held beliefs about the powers of the gene and the importance of heredity facilitate eugenic practices even in the absence of direct political control of reproduction. . . . These beliefs—conveyed through the stories told by popular culture—draw on the assumption that our social, political, and economic future will depend on controlling the genetic constitution of the species—the so-called human gene pool."[73]

When we ask those scholars who are teasing out the structure of the gene myth what's wrong with it and how we should combat it, we get the following. What's wrong with genetic determinism is that it locates the source of social achievements and social problems in the individual, and this tends to support conservative public policy that is pessimistic about the positive potential of social change. It tends to encourage genetic discrimination; it reinforces racism and classism and even patriarchy.[74] Science inadvertently serves the cause of social injustice. How, then, should we combat genetic determinism? The repeated answer is clear: add environmental determinism—that is, move from a single-cause determinism in the DNA to a double-cause determinism that combines genes and environment.

The dominant anti-gene-puppet proposal is to invoke nurture to supplement nature. Nelkin and Lindee contend that "in a trivial sense every human trait is biological. . . . At the same time, every human trait is environmental . . . many traits that are clearly genetic are also malleable, influenced by culture and environment."[75] In their diatribe, *Exploding the Gene Myth*, Ruth Hubbard and coauthor Elijah Wald make the same point bluntly.

> Many new genetic breakthroughs . . . do not make people healthier; they merely blame genes for conditions that have traditionally been thought to have societal, environmental, or psychological causes. . . . The myth of the all-powerful gene is based on flawed science that discounts the environmental context in which we and our genes exist. . . . What I object to is the reductionist effort to

explain living organisms in terms of the workings of important molecules and their component parts.[76]

In the recent struggle between nature and nurture, where nature has been gaining the upper hand, these critics of the gene myth want to rally nurture to their defense. Rather than see nurture on the retreat, they are calling for a counterattack against the army of genetic determinists.

What I find curious, of course, is that they are fighting one form of determinism with another form of determinism. There is no defense of freedom here.

Becoming Human

It is not my mission to go to war against the determinists in order to champion the cause of freedom. It is not my task to show how free will escapes the snares of biological and cultural conditioning to establish a *tertium quid,* some sort of third metaphysical entity known as the human spirit. I have proffered the more modest claim, of course, that the phenomenon of human freedom is a given. And I believe attempts to explain it as an illusion rely upon a category mistake, the reductionist mistake of confusing biological parts with the whole human organism. This means that genetic determinism—and cultural determinism as well, for that matter—are not incompatible with our everyday understanding of free will. If necessary, we could live philosophically with both genetic determinism and human freedom.

Nor is my mission simply to join forces with the army opposing the gene myth. Many of the critics of the gene myth whom we have reviewed are engaged in a cultural power struggle, an assault against the intellectual hegemony of the natural sciences. These critics want to do more than merely clarify what does and does not count as determining factors in human living; they want to rescue greater spoils of war for the ideology of social constructivism. The critique of genetic essentialism is simultaneously a critique of essentialism in general. The insights of these critics are extremely helpful, in my judgment. But I have no particular investment in who wins this battle.

Rather, my task is to reflect theologically on issues prompted by the gene myth with its maxim against playing God. The struggle over genetic determinism and human freedom is indicative of the ontological tension that Christian anthropology sees as inherent in the human condition. The fundamental struggle issues from the underlying tension between *soil* and *spirit.* We humans find ourselves at what Plato called the *metaxy,* the point at which we become aware of our own limitations while we dream of transcending them. We find ourselves conditioned and constrained by life on earth, yet we can dream of soaring to heaven. We can understand our genetic conditioning, yet we transcend it by envisioning freedom and then pursuing it.

The vocabulary of soil and spirit arises from the biblical story of creation, wherein God takes soil and fashions the first human being. Out of the dust of the earth God creates an inanimate form, then breathes divine breath into its nostrils. At this, the form becomes a living being. "The God the Sovereign One formed a human creature of dust from the ground, and breathed into the creature's nostrils the breath of life; and the human creature became a living being" (Genesis 2:7, ILL). The Hebrew word for breath is the same one for spirit, *ruach*. The Hebrew word for this first human creature is *adam,* from which we get the name Adam. The Hebrew word for the soil out of which Adam was made is *adamah.* *Adam* comes from adamah. The biblical point here is not that men come before women, otherwise Genesis would have used the male term for man, *ish,* in contrast to *ishah* for woman. Rather, Genesis used the term for all humankind, *adam,* inclusive of men, women, and children. The biblical point is that all of us in the human category of being are a combination of soil and spirit. We find ourselves in the metaxy, in a tension between belonging completely to the earth while at the same time breathing the breath of the divine. We find ourselves finite but dreaming of the infinite. We find ourselves slung between heaven and earth.

Critics of the gene myth are right when they refuse to identify DNA with the human soul. Our soul is not simply in the DNA. Nor is it in the culture. Our soul is not merely the product of environmental conditioning any more than a product of genetic determinants. Without positing its metaphysical status, I believe the human soul is a phenomenon that exists at the metaxy, at the interstice of soil and spirit. It is the human self, a locus of freedom racked with tension over what has been and what will be.

By the term *spirit* I refer to the dynamism of the human way of being, to our freedom and to our ability to create new futures that differ from our past. We order and reorder ourselves through deliberation and decision making. We see possibilities. We actualize some, but not all. We may have begun as soil, but by inhaling the divine spirit we now transcend our past and gain the freedom to create alternative futures.

This is the beauty and the tragedy of the soil–spirit tension. On the one hand, human vision and responsible action can enhance already existing freedom while making the world a better place. On the other hand, uninformed judgment or malicious intent can create evils that escape human control and wreak massive suffering.

Genetic determinism along with its partner, environmental determinism, belong to the soil. Our bio-cultural evolution constitutes the soil from which we the human race have been fashioned. Our evolutionary history is complex, and this past keeps us poised for influencing a possibly still more complex future. The presence of spirit in human consciousness draws us forward to new and unprecedented images of the future. Those images can include new pictures of what it means to be human. The understanding-decision-control structure of promethean determinism poises us for making tragic blunders; but it is itself indicative of the deeper metaxy that underlies the human condition.

Suffering, Death, and Resurrection

The tension between soil and spirit is not due merely to the tension between the finite and the infinite, between the conditioned present and the open future. It also reflects the conflict between evil and good. It reflects the tension between suffering and health, between injustice and justice, between bare survival and human flourishing. Even though the battle over the gene myth appears to be a secular one, below the surface it is ethical and even theological. More is at stake than simply discerning whether biological determinism has a stronger influence than environmental determinism. At stake here is the question: What will we do about human suffering? Should we pursue research into medical genetics to help relieve suffering, or should we temper if not abandon genetic medicine entirely because its advance may lead to social injustice and more suffering?

When we frame these questions within the conceptual set of the gene myth, particularly within the grand worldview of evolutionary psychology, the question of God's relation to the dynamic world of human evolution arises. The question of suffering within creation arises because the long history of natural selection is based not only on selective reproduction but also individual death and species extinction. Charles Darwin gave up on the idea of divine design in evolution because of the clumsy, wasteful, blundering, and cruel works of nature.

The human past is replete with suffering. So is the present. What about the future? If we speak about God as creator, and if we think of the evolutionary advance as the means of divine creation, then should we blame God for suffering?

Why is there suffering in this world? That suffering is built into natural selection, and the evolutionary process is simply accepted without question. Yet, for those of us who are born into this world with genetic defects and severe disabilities, the question arises: Why do I have to hurt so much?

In theological language the cry of suffering is taken up by the theodicy problem. Physicist and theologian John Polkinghorne puts the theodicy problem into these words: "How can a world of cancer and concentration camps be the creation of a God at once all-powerful and all-good?"[77] Or, we can frame it a bit more abstractly: How can we reconcile the following three propositions? First, God is omnipotent—that is, all-powerful. Second, God is omnibenevolent—that is, all-loving. Third, suffering exists. On the one hand, if God is omnipotent and suffering exists, then God must have created it and, thereby, cannot be all-loving. On the other hand, if God is omnibenevolent and suffering exists, then God must be too weak to prevent it. If we had a third hand, we could assert that God is omnipotent and omnibenevolent and then deny that suffering exists— that is, we could say that suffering belongs strictly to the physical world which is an illusory world compared to the much more real world of spirit beyond suffering. Christian theologians, no matter how difficult the challenge here, nearly always affirm all three propositions.

Note that we are working with two kinds of suffering here, some due to natural evil and some to moral evil. Natural evil and moral evil overlap, but they are distinguishable. Cancer belongs in the former category. Cancer like so many diseases is a product of nature, a horror created by the interaction of our genes with our environment. Every cancer is genetic at least in part, and some, such as inherited breast cancer, are explosions of a DNA time bomb placed within us at conception. The human suffering resulting from such a disease is nobody's fault. It is natural. It is an evil bequeathed to human individuals by nature. Many of us are brutally predetermined by nature to suffer. Theologians would add to cancer a long list of natural evils such as plagues, droughts, floods, earthquakes, tornadoes, and similar disasters that have taken untold numbers of human lives and left us with a devastating history of human suffering.

The second kind of suffering is due to human evil—that is, sin. Concentration camps produced suffering as the deliberate product of human hands. Child abuse, rape, street crime, organized crime, terrorism, torture, political totalitarianism, civil war, international war, and genocide are some of the ways that the human race has devised for drawing innocent people into a state of victimage. Suffering here is the product of human creativity. It is the product of human freedom. In contrast to natural evil wherein no human agency is directly responsible, this is dubbed moral evil or human evil. When advocates of the gene myth post "No Trespassing" signs against playing God with DNA, they are trying to protect us against possible moral evil.

How do we defend God in the face of either of these two forms of evil? The term "theodicy" means literally defending God as just. Quite frequently theologians employ the free-will defense. By asserting that God is self-limiting, they say God makes room for freedom in the world. And with freedom in the world, room is opened up for evil and suffering. The argument works best in the case of moral evil: the reason moral evil and human suffering exist is that God permits it as a price to pay when purchasing human freedom. Yet the argument can be extended to apply to the natural world as well. The limiting of God's determining power allegedly permits contingency in natural events; it permits evolution complete with its genetic mutations and natural selection to run its own history. It permits the created order to operate independently. And, if this independence results in suffering on the part of any creature, human or animal, then so be it. Polkinghorne provides us with a well articulated example of this position.

> This leads me to embrace the free-will defense: that despite the many disastrous choices (and one cannot say that in this century without a quiver in one's voice), a world of freely choosing beings is better than a world of perfectly programmed automata. . . . In relation to physical evil (disease and disaster) there is a parallel "free-process defense": that in his great act of creation, God allows the whole universe to be itself.[78]

Implicit here is an extremely high value placed on human freedom—freedom understood as independence from God's will and God's power—and, by analogy, the concept of human freedom is applied to the natural world. Suffering, whether natural or moral, is said to be the price God pays to purchase freedom at the cash register of the evolutionary process.

As for me, I am not sure I would want this price to be paid. I find the argument of God's self-limitation to permit suffering in order to attain human freedom less than fully satisfactory. I recognize that Christian theologians for two millennia have espoused variants of this position; and I am aware that esteemed contemporary theologians find it compatible with evolutionary theory. Yet, somehow I find it less than satisfying. Suffering, whether due to natural evil such as defective genes or moral evil such as concentration camps, I find too high a price to pay for what we have been describing as freedom. If the cosmic marketplace could make the offer, I would gladly exchange my freedom to put an immediate end to it. Conceptually I would find it easier to consign past evil to mere indeterminism, to contingency and chance, rather than to a divine plan.

I do not expect to solve the theodicy problem here. Nor do I expect to provide a complete anthropology that speaks to every aspect of evolutionary history. But I would like to invoke, for our understanding of human becoming, two items we have learned about God from the revelation in Jesus Christ. First, we have learned that God identifies with those who suffer and those who cause suffering. As Jesus spent time in solidarity with victims of social stigma—due to leprosy, being born blind, or other diseases—we believe that God incarnate has entered the realm of flesh and thereby experiences what we experience as soil, as natural evil. As Jesus spent time in solidarity with tax collectors, traitors, and prostitutes—sinners who were stigmatized by the righteous citizens—we believe that God incarnate has entered the realm of sin, and thereby God experiences what we experience as moral evil. Having set aside divine power and independence to enter the realm of finitude and sin, the cross reveals to us that God incarnate has become the victim of suffering rather than its permitter or perpetrator.

Second, we can look forward to transformation. The death of Jesus was followed by his resurrection. The present creation will be followed by a new creation. Beyond suffering and death God brings new life and healing. In the cross of Christ, God identifies with the victims of injustice and suffering. In the resurrection of Christ, God promises us that injustice and suffering will not be in the last chapter of the human story. What this means for us now while the human story is still being written is this: we need not accept present reality as final. God's promise for a new future translates into future freedom and neighbor love for us.

The decisive element in a picture of future humanity—the picture as drawn by a Christian theologian—is resurrection. As Jesus Christ rose from the dead on Easter, the theologian remembers God's promise that we too shall rise into the eschatological kingdom of God. This is a promise for our future. Essential here is that resurrection is a divine act. God raised the dead Jesus. We will need God to raise us as well. Nothing in the DNA will prevent our dying, nor will DNA see us through to some form of postmortem existence. Resurrection clearly transcends genetic and cultural determinism. Yet, from the Christian point of view, resurrection belongs to true humanity. Resurrection is essential to the ontology of human nature.

There is definite continuity between our DNA and our destiny. The future will not nullify our past or present. Our spirit will not escape our biology. The new creation will fulfill, not replace, the old creation. Yet we can expect something new, an epigenesis that will take us beyond where our evolutionary history has brought us thus far. Resurrection and eternal life will become natural. This is God's promise. This means, theologically

speaking, that genetic determinism and environmental determinism combined have not yet brought us to our full humanity. We are still on the
way, so to speak. Living in the tension between soil and spirit makes us
restless, and in this restlessness we can hear the faint call of God to look
forward to a future that will transcend our long and complex past. Our
resurrection will be a future that only God can create. In the meantime,
life at the metaxy can in its own way be quite creative as well. Poised for
beauty and tragedy, we have a growing future freedom that is calling us to
responsibility.

< THREE >

The Crime Gene, Stigma, and Original Sin

> Our attempts to grasp the meaning of our freedom have constantly
> been blocked by a guilty sense that we must not tamper with the
> machine-picture because it is stamped as final by Science.
>
> —Mary Midgley

Can I legitimately say the following to the judge? "Your Honor, I'm innocent. My genes made me do it." Might the gene myth play this kind of
role in the courtrooms of the future?

Glenda Sue Caldwell's case may be the advance ticket for the show that
is to come. On July 7, 1985, as her son Freeman walked through the front
door of her house, she shot him three times. Freeman died. Mrs. Caldwell
took this gun and another gun into the bedroom where her daughter,
Susan, was sleeping. She fired at Susan, missing but leaving powder burns
on Susan's face. On June 4, 1986, she was convicted of murder and
assault and condemned to a life term and a ten-year concurrent sentence
in prison.

Her defense? Innocent on grounds of diminished capacity due to insanity. The insanity was said to have been brought about by her fear of con-
tracting Huntington's Chorea disease combined with the stress of
separation from her husband. The insanity defense currently requires that
one of two criteria be met: either the defendant does not have the ability
to distinguish right from wrong, or else the defendant is acting under a
delusional compulsion that overmasters the will to resist committing
crime. The Georgia jury in the Superior Court of Clayton County,
presided over by Judge Kenneth Kilpatrick, was unpersuaded by the insan-

ity plea and found Mrs. Caldwell guilty. The guilty verdict was affirmed April 8, 1987, by the Georgia Supreme Court, wherein Justice Hunt opined that Mrs. Caldwell knew right from wrong, was not a victim of delusional compulsion, and was attempting to hurt her estranged husband.[1]

Case closed? No. On August 25, 1994, Judge Kilpatrick without a jury retried Mrs. Caldwell, declared her innocent by reason of insanity, and set her free.[2] While serving nine years in prison, symptoms of Huntington's disease had begun to manifest themselves. Huntington's is a terminal brain disorder that degenerates nerve clusters in the brain, causing involuntary movements and dementia. Symptoms include depression, impulsiveness, poor judgment, and sometimes violent behavior.[3] Mrs. Caldwell's symptoms included a brain tumor which was removed surgically while in prison.

Key here is that the predisposition to Huntington's has been traced to a particular gene on chromosome 4. It is an inherited disease. Mrs. Caldwell's father and brother died from it. She was seeing herself as genetically condemned when she pulled the lethal trigger. Nearly a decade later this inescapable genetically determined fate became the grounds for establishing the new not-guilty verdict. In effect the court ruled: Mrs. Caldwell is not responsible for her violent behavior; her genes are.[4]

"I always knew that something was wrong with me," Glenda Sue Caldwell told reporters upon her release. "I was not responsible for what I did. I'm a good person."[5]

As I write this, I have no desire to add any more grief to the sadness already experienced in the Caldwell family. Rather, I would like to keep the focus on what might serve as a precedent regarding how our justice system might handle the application of genetic determinism to moral responsibility. A curious paradox is in the offing. On the one hand, we can foresee an assumed but unarticulated doctrine of naturalism—naturalism says that if it's natural, it's good—utilized to label as innocent persons who engage in immoral or dangerous activities. On the other hand, we can foresee the rise of new stigmas, new patterns of discrimination against persons known to be carriers of an undesirable gene.

Perhaps we should note that there are approximately 25,000 sufferers from Huntington's disease in the United States. How many of the other 24,999 have murdered their children? Have they been able to maintain responsible lives despite their genes? Victims of this disease would be appalled to think that their malady might become associated with crime and elicit public fear. The Caldwell case risks stigmatizing Huntington's sufferers as violent. At least this is the worry of Philip Cohen, a New York spokesperson for the Huntington's Disease Society of America. "Not every Huntington's person is violent. That would be a gross mischaracterization."[6] One possible scenario from such a precedent is that genetic determinism might end up declaring those committing crimes innocent and stigmatizing those not committing crimes as potentially guilty.

When it comes to antisocial behavior such as murder and stigmatization, the interpretive filter of the gene myth should give us pause to inquire about its ramifications. Does genetic influence on behavior simply bypass the human self or human will when coming to expression? If so, does this remove moral responsibility from us as persons, or does it increase the weight of the load of moral responsibility? Does this newly discovered biological propensity for evil indicate that we now have scientific confirmation of the ancient doctrine of original sin?[7] Or, is it an argument for original innocence? If we use genomes to identify various groups of potential sinners, could we then justify social discrimination against them? Should we establish a program of mass genetic screening of newborns so we can identify those with criminal genes; and should we then send them immediately to prison, thereby saving society the grief of having to arrest them after the fact? As long as the gene myth continues to delete the concept of the self, to banish from our thinking the human person with a free will and social responsibility, we will come to fear the genes and ignore the person.

A Gene for Violence?

There is good reason for molecular biologists to search for genetic predispositions to behavior in general and to antisocial behavior in particular. Genes have a decisive influence on our phenotype and on the daily operations of our bodies. It simply stands to reason that many—perhaps all—propensities for human acting could have genetic origins. This would be no less true for socially undesirable behavior as it would for any other behavior.

Gene hunters do occasionally make dramatic finds. For example, in 1993 a team of researchers in Holland discovered a genetic marker on the X chromosome that identifies a predisposition to violence. Studying the genomes of an extended family wherein eight of the men but none of the women had clearly identifiable antisocial behavior, they located a common DNA sequence that distinguished them. All eight of the affected men were mildly retarded, a typical IQ score at 85. Characteristic of their behavior was exhibitionism and voyeurism. They engaged in repeated episodes of aggressive violence, usually triggered by anger that was out of proportion to the provocation. One who had raped his sister was incarcerated, and while on a prison farm, he stabbed a warden in the chest with a pitchfork. Another affected member of the family tried to run over his boss with a car. A third would enter his sisters' bedrooms at night, armed with a knife, and force them to undress. At least two of the men are known arsonists.[8]

This does not by any means confirm common suspicions that all men have a gene for aggression any more than they have a gene for watching the Super Bowl. The study identifies one group of eight men. The significance is that a correspondence seems to have been discovered between a gene or combination of genes and a specific form of behavior. Could there be many more such genes in the human genome?

Suppose there are. Suppose we find a number of genes or configurations of genes that elicit attitudes or stimulate states of mind that incline us toward behavior that is harmful toward others. If we remind ourselves of what molecular biologists said regarding the relation between the genotype and the phenotype and the present lack of threat from genetic predestination, then it would be premature to think in terms of excusing criminal behavior on the basis of DNA. This is certainly the judgment of law theorist Maureen P. Coffey. After examining the existing court precedents, Coffey concludes that "recognition of biological determinism need not require the adoption of a constitutional or special genetic defense. Because an individual may be *more vulnerable* to developing a chemical addiction or an antisocial personality disorder does not mean that the individual *in fact* will develop those conditions, or that the individual has absolutely *no control* over such development. The existence of a genetic condition merely provides more insight into whether a person possessed enough free will or rational ability to control and understand her behavior."[9]

What if in the future we could demonstrate scientifically that "it's all in the genes"? What if the gene myth proves true? Then Coffey would argue for incarceration on the grounds that society needs to be protected from criminal behavior. "While incarceration of a genetically afflicted offender would not serve the traditional objectives of retribution and deterrence, society nonetheless may determine that the isolation function outweighs these objectives."[10] That is, the court could decide to restrict the social freedom of a person if his or her genome identifies a threat to society.

That society needs to be protected from criminal behavior, and that such protection could be had by isolating persons with certain genetic dispositions, leads to further questions regarding insanity and race. The issue of insanity arises because we can predict that the genetic defense may rely upon precedents set by the insanity defense. The courts treat insanity with a focus on the insane person's inability to distinguish right from wrong when committing a crime. When a defendant is judged innocent on these grounds, he or she is incarcerated in a mental hospital until the medical evaluators judge that the individual is cured. Once cured, the patient is released. In principle, such a person might never be judged "cured" and spend more time in isolation than the prison penalty prescribed for the crime, maybe even the rest of his or her life. Should the genetic defense tie itself to the insanity defense, and if one's DNA is thought to last a lifetime, then the trip to the hospital may become the equivalent of a life sentence. In this way the genetic defense may backfire. When it backfires, it may be picked up by the prosecution for the purpose enunciated by Coffey, namely, to incarcerate certain individuals with certain genomes.

This is a terrifying thought. Coffey, of course, is merely drawing out the terrifying implications of the puppet form of genetic determinism. One implication would be stigma, and those stigmatized could find themselves incarcerated.

Is There a Crime Chromosome?

Instructive in this regard is the story of XYY trisomy. College biology textbooks routinely report that men with 47 rather than the normal 46 chromosomes—the extra one being an extra Y chromosome—are slightly taller, mentally retarded, and have hormone imbalances, childhood reading disabilities, outbursts of anger, and, most importantly, aggressive anti-social behavior. Why does this get reported in those textbooks?

The first recorded discovery of a man with the XYY configuration appears in 1961 in England; the man had a history of barroom brawling during his youth. This led to a genetic study of prisoners and a 1965 journal article in *Nature* that, like a bull in a china shop, has left debris scattered over the decades since. A study of the karyotypes (chromosomal complement) of 197 men, all patients in Carstairs, a high-security mental hospital in Scotland, showed what the authors thought to be a disproportionately large percentage of XYY inmates. Seven of those studied carried the XYY trisomy, approximately 3.5 percent of the total. They were described as "mentally subnormal male patients with dangerous, violent, or criminal propensities."[11]

The *Nature* authors speculated that this 3.5 percent may be twenty times higher than men in the normal population. However, no one in 1965 knew how many men in the normal population are XYY; so no control existed to support the speculation that the XYY genotype might be an influence on imprisonable behavior. Similar studies in the decade that followed similarly neglected to include as subjects XYY men who were free and contributing normally to society.

Like leaflets dropped from an airplane, irretrievable and unrevisable news spread that XYY babies were doomed to grow up as social problems. DNA had determined their destiny. "Congenital Criminals" was the title of a *Newsweek* story. The drama increased when newspapers in 1968 announced that Richard Speck, the notorious murderer of eight student nurses in Chicago, was XYY. The report was false. Speck was XY. But it was too late to modify the public image: if XYY, then violent.

These men became known as "super males." A certain common-sense logic led to this. The Y chromosome is a sex-determining chromosome. Women have two Xs. Men have an X and a Y. By and large men tend to be more aggressive than women. Could this be due to the presence of the Y chromosome? Perhaps. Now, suppose we add an extra Y. Would this add an increment of aggressiveness? Do XYY men have a double dose of masculinity? Such has been the reasoning. Is it accurate?

Enter abortion. As we proceeded through the 1970s maternity hospitals in England, the United States, Canada, and Denmark began permitting mass screening of newborn infants to identify those with XYY genotypes. Some scientists proposed that genetic testing be done prenatally. Since no remedy for this malady existed, an implied purpose of the proposal was to support selective abortion. Informing parents that their future baby boy would carry the trisomy would give them the option of terminating the pregnancy.

The idea of aborting XYY males may on the surface sound like good eugenic policy. If this genotype is a heritable trait, and if we prevent such people from being born and later passing this trait on to subsequent generations, then eventually we could eliminate XYY from the general population. This would make the world a less aggressive and safer place. However, such a policy would not work. Why?

One male baby in a thousand is born with XYY trisomy. But the cause is not the genotype of the parents. The cause is a nondisjunction that takes place in the paternal reproductive cells during meiotic metaphase. Simply put, two Y chromosomes sometimes get stuck together and fail to separate during cell division. Like Siamese twins, they travel together in the sperm. When this sperm fertilizes an egg with an X chromosome, the result is a male XYY embryo. Because the process happens spontaneously, no eugenics program could prevent XYY boys being conceived even if selective abortion would prevent them from being born.

What about the legal logic of XYY? If XYY constitutes a genetic orientation toward crime, could we ask the courts to blame the chromosome and not the person commiting the crime? Could we ask for exculpation? Savvy lawyers have occasionally tried making such genetic defenses. In general the courts have been unsympathetic to the chromosome defense—usually a version of diminished capacity by reason of insanity defense—but they have left the door open should confidence in the scientific veracity of the claims be strengthened. In *People v. Tanner*, for example, a California appellate court upheld a lower-court ruling barring the use of the XYY defense on the grounds that the link between XYY and insanity was not "clear and convincing."[12] What the courts want to see proven is that the XYY syndrome consists in a biological process that so interferes with the defendant's cognitive capacity that he or she is unable to choose freely between right and wrong. To date this is unproven to the satisfaction of the legal community, as it is unproven in the scientific community.

By the mid 1970s the accepted notion of the criminal chromosome was under scientific attack. The original study and those that followed were sharply criticized for leaving out comparative samples of XYY men in the general population. The studies were said to be flawed because they relied upon simplistic and vague definitions of human behaviors that contribute to social deviance or crime. They also were criticized for ignoring or downplaying environmental factors in criminal development such as family history, nutritional deficiency, economic class, and such.[13] Although the subject matter connecting genes and crime may very well be worth studying, the XYY plank in the gene-myth platform is sufficiently warped as to recommend its removal. Anthropologist Stephen Jay Gould contends that while "a maximum of 1 percent of XYY males in America may spend part of their lives in mental-penal institutions . . . 96 percent of XYY males will lead ordinary lives and never come to the attention of penal authorities. Quite a criminal chromosome! Moreover, we have no evidence that the relatively high proportion of XYY's in mental-penal institutions has anything to do with high levels of innate aggressivity."[14]

The ethical danger with this chapter in the history of genetics is stigmatization. David Suzuki and Peter Knudtson, authors of a widely read textbook on genethics, regret the hasty hypothetical association between chromosome and character. "Once a young boy . . . was found to possess an XYY genotype, he could be marked for life—burdened with an unearned scarlet letter in the form of a second, or supernumerary, Y."[15] These two go on to attack the gene myth. "We run the risk of investing readily detectable hereditary traits with almost mystical prophetic powers. Like some prescientific shamanistic healer who finds people's fortunes revealed in fallen strands of hair or the discarded clippings of their fingernails, we unconsciously begin to extrapolate from the qualities of the part the qualities of the entire organism."[16]

The Suzuki and Knudtson comment deserves further comment from us. By recognizing how the gene myth extrapolates qualities belonging to the organism as a whole—in this case, the qualities of aggressiveness and criminality—and applies them to the part—in this case, the extra Y chromosome—these two are presupposing holism. They are on the border of factoring into their analysis the human person with a free will who makes decisions. But they stop short of crossing the border. In order to combat genetic determinism they take the default position of two-factor determinism, genes and environment. This simplistic move is hidden within this otherwise nuanced observation: "Most geneticists agree that the vast majority of human behavioral traits tend to be polygenetic; they reflect the simultaneous interaction of a multitude of genes. And every element in our species' rich behavioral repertoire is unavoidably sculpted, at every stage of its development and action, by a myriad of environmental variables too numerous to possibly anticipate."[17] There is no defense of freedom here, just a camouflaging of two-part determinism in a barrage of uncountable determining factors.

Some strategists in this battle try to elicit empathy for the XYY males among us on the grounds that each of us, regardless of the number of our sex chromosomes, carry some undesirable genes. The suggestion here is that we are all in the same boat, so none should get pushed overboard. Bruce Reichenbach and Elving Anderson, for example, note that "it is estimated that every person carries five to ten seriously defective genes . . . the revelation of our own genetic fallibility can be the occasion either for acceptance of the human condition of finitude or for guilt and self-incrimination."[18] This argument is charitably minded, to be sure; but it is not likely to carry much force. This is because it neglects one important element, namely, the fear of harm that violent behavior produces. If we carry a defective gene for breast cancer, then we certainly can feel empathy for someone else who carries the gene for cystic fibrosis. But if that other person carries a gene predisposing him or her to harm us, then this adds an additional element of considerable consequence.

Genes, Crime, and Race

Fear of stigma rides genetic determinism like a jockey rides a horse. Patricia A. King alerts us, "The danger to racial and ethnic minorities and the poor from current gene mapping efforts is obvious: the danger is that greater attention will be paid to genetic explanations than to more complex explanations for differences to the detriment of vulnerable and disadvantaged groups."[19]

In 1992 a major controversy broke out over the relationship between genes and crime. David Wasserman at the University of Maryland had secured a grant from the Ethical, Legal, and Social Issues (ELSI) program at the National Institutes of Health (NIH) to fund a conference he had planned on "Genetic Factors in Crime: Findings, Uses, and Implications."[20] Professor Wasserman had planned an open academic discussion of the scientific findings, findings that would most likely have shown that to date no serious biological evidence exists directly linking genes to crime. This conference did not take place as scheduled. The brochure advertising the conference was misleadingly inflammatory, and a public protest arose.[21] A black interest magazine announced, "U.S. government wants to sedate black youth" in order to counteract their genetic propensity to crime. Ronald W. Walters, a Howard University political scientist, announced opposition to all research into the biological causes of crime. The nation's capitol was bombarded with complaints. The Black Caucus in the U.S. Congress responded. The government, sensitive to the public outcry, capitulated. Bernadine Healy, then director of the NIH, demanded that funding be revoked. NIH Deputy Director John Diggs canceled the conference.

There is disagreement over whether Dr. Healy did the right thing. Among those who approve of Healy's action are Ruth Hubbard and Elijah Wald, who accuse Wasserman of drawing "attention away from the societal reasons for why poor people, especially African-Americans, make up a disproportionate share of America's prison population." They add that "the very definition of criminal behavior is flawed, and that it is meaningless to talk about genetic predispositions for behaviors that are socially constructed." Then they conclude by saying that "there is no scientific merit to looking for genetic factors in crime."[22]

Paul Billings, a geneticist with special concern for ethical issues, takes the opposing view. Along with Hubbard and Wald, Billings also doubts that scientists will find a direct link between genes and crime. Nevertheless, he disagrees with the decision to cancel on the grounds that "the NIH did a disservice to academic freedom and allowed the fantasy that there is a science behind the discussion of the genetics of criminality to continue."[23] In sum, the scientific consensus seems to be that genes and crime have at best an indirect connection, and some scholars are anxious that drawing premature conclusions regarding genetic causes of crime might have deleterious repercussions for race relations.

Berkeley sociologist Troy Duster is worried about such racial repercussions. If we identify crime with genes and then genes with race, then we may inadvertently provide a biological support for prejudice and discrimination. He sounds the alarm: "Today, the United States is heading down a road of parallel false precision in this faith in the connection between genes and social outcomes. This is being played out on a stage with converging preoccupations and tangled webs that interlace crime, race, and genetic explanations."[24]

Wasserman, who describes himself as "a liberal academic and criminal defense lawyer," is sensitive to the possibility of racial stigma while still defending academic freedom and our responsibility to treat matters such as this scientifically. Even though opponents of genetic research into crime and violence argue that it promotes racism and diverts resources from social programs, Wasserman says the facts do not warrant these accusations. "No mainstream researchers believe that there are single genes that cause violent or antisocial conduct;" he writes, "all regard behavioral phenotypes like criminal behavior as arising from a complex interaction of many genes and environmental factors." He then raises the problem of stigma, worrying about understanding and empathy directed toward persons with certain genotypes. Most troubling, he finds, is that

> research on genetic predispositions may discourage any attempt to understand and empathize with those individuals *of all races* who are violent and predatory. . . . These efforts at understanding and empathy need not make us more lenient: we may still hold the offenders accountable for what they have done, and find their actions deserving of blame or reproach. But we resist the wholesale moral abandonment reflected, for example, in legislative efforts to consign repeat offenders to permanent, irrevocable imprisonment.[25]

Like Duster, Wasserman fears that the medicalization of the crime problem may lead society to treat violence-prone persons as dangerous beasts rather than as morally responsible agents, to be drugged or confined rather than educated or reproached.

Speculative Genetics

The track on which we have been running with the criminal chromosome has thus far been molecular biology. As we now turn away from physical science to more speculative genetics, the same axel with its two wheels of genetic determinism and stigma continues to roll on.

James Q. Wilson, a professor of government, and the late Richard J. Herrnstein in experimental psychology, both at Harvard, have been looking to nature to explain crime. They and their followers are looking at our biological or "constitutional" nature. They begin by observing that some features of crime are universal. For example, all societies and all classes agree that murder, theft, robbery, and rape are crimes. Different groups may differ regarding whether or not abortion or homosexuality is criminal, but the overwhelming majority of people of all locations and all classes agree that certain acts are unqualifiedly wrong and subject to social sanction.

Criminal activity, Wilson and Herrnstein further note, is greatest among young urban males. This applies to cities everywhere in the world regardless of race or class. In addition, the largest percentage of these urban males already exhibited behavior consistent with criminality at a very early age. Crime is also found in rural areas, to be sure; and it is also the case that some women are occasionally convicted of crime; nevertheless, a distinct set of urban men exhibiting lifelong criminality can be discerned. It seems reasonable to these two Harvard professors to ask whether there may be some constitutional factors that account for the relative prevalence of urban male crime.

Wilson and Herrnstein found that certain physical features can be associated with criminality. Criminals are more likely than noncriminals to have mesomorphic body types. They are more likely to have biological fathers who were criminals; and this applies even to cases where sons were adopted away and never even knew their birth fathers. They also tend to be of somewhat lower intelligence, to be impulsive or extroverted, and to have autonomic nervous systems that respond more slowly and less vigorously to stimuli.[26] This is the list of constitutional factors that dispose some individuals toward socially unacceptable behavior. These biological predispositions by themselves, of course, do not determine whether or not a given individual will actually engage in criminal activity. To commit a crime or not to commit a crime is a matter of a person's choice, and each person makes this choice on the basis of anticipated rewards or punishments. If the reward seems greater and will come sooner as the result of a crime, then he may choose it. If the reward for noncriminal behavior seems more attractive, then he will resist the temptation. In other words, physical factors, not merely social factors, influence our perceptions regarding reward or punishment.

Is this strict biological determinism? No. As with so many who oppose strict genetic determinism, they affirm a dialectical or interactive relationship between nature and nurture, between physical constitution and surrounding environment. They assert that "there *is* a human nature that develops in intimate settings out of a complex interaction of constitutional and social factors, and that this nature affects how people choose between the consequences of crime and its alternatives."[27] They add: "There is no 'crime gene' and so there is no such thing as a 'born criminal,' but some traits that are to a degree heritable, such as intelligence and temperament, affect to some extent the likelihood that individuals will engage in criminal activities."[28] Despite these disclaimers and the embracing of the interactive view, however, what we have here is a strong emphasis on predispositions toward crime that some of us inherit. Some sons in particular inherit biological temperaments conducive to criminality from their fathers.

This observation of the genetics of young urban males leads Troy Duster to caution us about the risk of stigma. Duster is not denying what molecular biologists might find regarding the power of the genotype to influence the phenotype. Rather, speculative genetics that seeks to medicalize social problems worry him. He is worried that we might jump prematurely to conclusions regarding the social import of such findings. Duster states:

> In the last few decades, gradually, almost imperceptibly, our thinking about human social life has shifted to accept a greater role for genetics. . . . The last decade has seen a geometric increase in publications pronouncing the genetic basis of such disparate phenomena as shyness, rape, mental illness, alcoholism, crime, even social and economic position. How did we get here? . . . I will argue that the social concerns of an age, not the scientific status of the new knowledge-structure of genetics, offer the most compelling answer to this question.[29]

The social forces governing the interpretation of scientific findings are more influential than the findings themselves, Duster thinks. So he is worried about the role that an apparently scientifically derived image of criminality might play in our society.

Cause for worry arises from the observation that from 1981 to 1991 in the United States the prison population grew from 330,000 inmates to 804,000. He further notes that the vast majority of the new inmates are black. In 1991 African Americans were being incarcerated at a rate seven times higher than that of Caucasian Americans. Fifteen percent of black males between 16 and 19 years old will be arrested in any given year. With these facts in mind Duster asks: Might there come a day when researchers taking cell samples from cell inmates begin to draw conclusions that one race is more genetically disposed to crime than another race? And what would be wrong with this? Duster points out what we all know, namely, there is a difference between committing a crime and becoming convicted of a crime. Only a small percentage of those who commit crimes are arrested, still a smaller number are prosecuted, and even a smaller portion are incarcerated. Duster, a social scientist, is well aware that incarceration rates are a function of social, economic, and political forces. With this in mind, he sees the threat as this: methodologically biased science may lead our society to think they have a genetic explanation for why African Americans allegedly commit more crimes than others, but such an explanation would only constitute one more expression of an as yet unexpurgated racism. Applying genetic determinism as an explanation for crime risks an illegitimate scientific absolution that would obscure or even hide our massive social justice problem regarding race relations.

Intellectual honesty, of course, requires that we push on with laboratory research to see just how much and how little genes influence who we are. Still, we dare not permit ourselves to be lulled into a scientific pangloss that diverts our attention away from moral responsibility for seeking social justice.

Alcoholism: Is It in the Genes?

What about the connection between crime and alcoholism? The statistical correlation between alcohol use and crime is overwhelming. Fourteen studies have shown that alcohol plays an influential role in 60 percent of all murders and 40 percent of rapes.[30] Should we say, then, that alcohol causes crime? No, at least not to Wilson and Herrnstein. They cautiously suggest that alcoholism overlaps with one kind of criminal, the neurotic kind. Extroverted aggressive criminals, in contrast, may not be alcoholics even if they drink frequently. The largest nuisance, the drunk who causes traffic accidents, seems to have no relation to crime. So the two Harvard researchers conclude that "Alcoholics tend to be neurotic personalities; overtly aggressive men tend to be impulsive ones. Both alcoholism and aggressiveness may have some genetic component, but, if it is a different component, there would be no reason to assume that a predisposition to alcoholism and a predisposition to criminality would appear in the same person more frequently than by chance."[31]

With this in mind, let us ask about the possibility that susceptibility to alcoholism might be genetically inherited. Quite possibly yes, answers psychiatrist C. Robert Cloninger. Cloninger distinguishes two types of alcoholism. Type I refers to mild and late onset alcoholism that seems to derive from habitual drinking. The precipitating cause is clearly environmental—that is, alcoholism results from the consumption of strong drink over a long period of time. This is the most common and least severe. In addition there is a smaller group, perhaps one fourth of all alcoholics, that seem predetermined to alcohol addiction. These Type II alcoholics begin drinking to drunkenness as teenagers, find it difficult to maintain a job, and have troubled family lives. In fact, it is a family affair because the susceptibility seems to be inherited. Type II men seem to outnumber Type II women by a ratio of four or five to one and, like criminality, boys inherit this propensity from their fathers.[32]

Cloninger identifies three personality traits associated with Type II alcoholism, personality traits that appear to have a physiological basis. The first trait is novelty-seeking, which is "a heritable tendency toward frequent exploratory activity and intense exhilaration in response to novel or appetitive stimuli."[33] What happens is that the chemical dopamine in the nervous system is stimulated by alcohol and the person gets a pleasant feeling: a reward. Experiments show that sons of alcoholics experience a much stronger response to dopamine than those of nonalcoholics. Second, alcohol tends to block the serotonin system—that is, it tends to block the system of feelings that includes fear, inhibition, shyness, and such. Type II alcoholics are born with a rather sluggish serotonin system and are comparatively less inhibited and more carefree. Third, Type II persons are born with a low sensitivity to norepinephrine, the neurotransmitter for signals announcing possible rewards and punishments. Type II alcoholics are less subject to behavioral governance by social approval. In sum, Type I alcoholics have high norepinephrine and serotonin sensitivity, low dopamine sensitivity, and are more susceptible to environmental influences; Type II alcoholics inherit a nervous system that has high sensitivity to dopamine, which is thought to lead the individual to seek novelty, while low serotinin and norepinephrine sensitivities lead to lack of inhibition or subjection to control by the routine system of social rewards and punishments.

Might Type II alcoholism be genetic in the molecular sense?[34] Some think that the gene responsible for the dopamine receptor on nerve cells can be found on chromosome 11. Is this enough evidence? No. More than likely it would take more than a single gene to dispose a person, perhaps a configuration of a dozen or more contributing genes. To date the Cloninger research has been suggestive but not decisive. It has not been independently confirmed. The scientific jury is still out.

Even though the scientific jury is still out, what about real juries? How might the alcohol gene combined with the crime chromosome play in court? If it can be demonstrated persuasively that a propensity to alcoholism or even to crime is genetically determined, will this count against or in favor of one's moral responsibility? Will genes become a scapegoat, permitting us to claim innocence by saying, "the genes—not the Devil—made me do it!" Or, will we think just the opposite, that our genes make us guilty? It could go either way.

Fascinating in this regard is the 1989 California case, *Baker v. State Bar*.[35] Here Mr. Baker was accused of embezzlement. The accusation of embezzlement went uncontested. His defense consisted of asking the court for leniency on the ground that he "had a genetic predisposition to alcoholism." He committed the crime under the influence of alcohol. The court seemed persuaded by this argument and, though convicting him, mitigated his sentence. He was not exculpated of the crime; his sentence was lessened. The ground for mitigation was that Baker had not known of his genetic predisposition toward alcoholism when, under the influence, he stole money. Now, following the hearing, he possesses this knowledge about himself. This implies that he now has the cognitive ability to control his alcoholism and avoid future criminal acts. If he returns to court under a similar charge, he will receive the full sentence. The assumption of the court is significant: Baker cannot be reduced to his genes. As a person, as a self, he possesses sufficient freedom regarding his own behavior that society can expect him to be responsible.

This leads to an interesting philosophical concern. What role should knowledge of one's genome play? Why is ignorance of one's genetic predisposition mitigating, but knowledge of it is not? If the gene myth is at work in the court so that we are assuming puppet determinism, then why might we assume that knowledge of such determinism itself is liberating? On what grounds do we cede the power to regain self-control to knowledge about our genotype ?

Regardless of the distinction between knowing or not knowing one's own genome, sociologist Dorothy Nelkin finds the mitigation of Baker's sentence alarming. As we have seen, she calls the gene myth "genetic essentialism." This essentialist position, if it spreads, could mean that all those proven to be biologically wired for alcoholism or criminality or both could use their genetic inheritance as a defense. To be biologically determined is to be innocent. If essentialism is set loose on the court system, and if Wilson and Herrnstein are correct, then the vast majority of predatory criminals will enjoy freedom from incarceration and our society will undergo unparalleled social anarchy.

Of course, society might consider the alternative, namely, assigning guilt to persons due to genetic inheritance. But this also has pitfalls. If genetic predisposition is used to support extending rather than mitigating punishment, and if society determines that persons so genetically predisposed should not be set free to create anarchy, then we might find ourselves in a situation similar to that created by the insanity defense. In some cases where the defendant claims diminished capacity or insanity, the court will remove him or her from society for a period longer than the maximum criminal sentence otherwise prescribed for the offense in question. In the future courts may decide to base sentences on predictions of dangerousness, and such predictions of dangerousness could be based on a genetic calculus. Genetic endowment would create a sort of status offense, and this may very well open the door to racial or ethnic prejudice. As Nelkin puts it, "If it is accepted that genetic endowment determines the propensity to commit bad acts, then hereditary traits, which often reduce to ethnic group membership, may one day be considered evidence of the commission of a crime."[36] Fears of stigma would be justified.

The Ethical Clang of the Bell Curve

Amidst the din of warfare between nature and nurture, the forces of the gene myth have rung out a clarion anticipating a decisive victory for the genes. The secret genetic weapon thought to assure eventual victory is IQ. This is the thesis of the 1994 best-selling book, *The Bell Curve*. Filled with recommendations for social policies based on IQ that may perpetuate existing racial and ethnic stigmas, this book has provoked alarm over the stigmatizing power of the gene myth.

For *The Bell Curve*, coauthor Richard Herrnstein teamed up with political scientist Charles Murray. These two argue that intelligence, understood as cognitive ability, is inherited and is decisive in determining a person's life-chances for success in education, income, class status, and contribution to society. They hold as "beyond significant technical dispute" such propositions as: cognitive ability exists and differentiates some human beings from others; IQ tests measure cognitive ability with sufficient accuracy; high IQ scores match with what we mean by *intelligence* or being *smart*; properly administered IQ tests are not demonstrably biased against social, economic, ethnic, or racial groups; and cognitive ability is substantially heritable, apparently no less than 40 percent and perhaps up to 80 percent.[37] The central thesis is that "cognitive ability is the decisive dividing force" between social classes. "Social class remains the vehicle of social life, but intelligence now pulls the train."[38]

To support these contentions, Herrnstein and Murray use statistics gained from IQ testing, such as data gained from the National Longitudinal Survey of Labor Market Experience of Youth (NLSY). They do not rely upon knowledge produced by molecular biology, nor do they appeal to behavioral genetics or evolutionary psychology. They engage in no testing of their own. Their method is to offer an interpretation of existing statistical data originally gathered for other purposes. Such a method is certainly adequate if properly qualified; but it ought not to be considered original science.

This version of genetic research finds its place within a social agenda that begins with the premise that the quality of American life is deteriorating. This deterioration is allegedly being caused by the welfare state— that is, by a government policy that provides too much social and economic support for those among us with lower cognitive ability. Herrnstein and Murray want to reverse the welfare trend. They argue that low IQ is generally associated with socially undesirable behaviors— poverty, school dropout rates, unemployment, job-related injuries, illegitimacy, divorce, welfare dependency, and crime rates—and this association warrants withdrawal of public support for groups of people known for their lower cognitive abilities.

"High cognitive ability is generally associated with socially desirable behaviors," they write, and "low cognitive ability with socially undesirable ones."[39] This applies to groups, not individuals. It applies to social or ethnic classes, not to particular persons, they say. We cannot predict what an individual person will do on the basis of his or her IQ score. Yet statistically we can identify large differences in social behavior between separate groups of people when the groups differ intellectually on the average. With this observation in mind, Herrnstein and Murray contend that it is intelligence itself, not just its correlation with socioeconomic status, that is responsible for these group differences.

What are these average differences? In ranking by race, African Americans come out lowest with an average Intelligence Quotient of 85. Next up the ladder we find white people, averaging about 100. Still higher up the ladder are East Asians (Chinese and Japanese) who exceed Euro-American whites by 3 to 10 points. On the very top of the ladder are Ashkenazi Jews of European origin, who test higher than any other group. Regardless of level, men and women have nearly the same mean IQs, even though the men have a broader distribution—that is, proportionately more men are found at the extremes of low and high IQ. All this "suggests, without quite proving, genetic roots."[40]

Given the premise that IQ and desirable social behavior are "generally associated," we would expect Herrnstein and Murray to lift up East Asian and Ashkenazi Jewish behavior as models for all the rest of us to celebrate and emulate. But this is not the program they have in mind. Rather, they propose a form of economic eugenics aimed at reducing the number of children born to unmarried African American women and lower-class white women. This prescription follows upon their description of what is happening in America, namely, the cognitive elite is increasingly finding one another and establishing an intellectual-entrepreneurial meritocracy that will eventually place the nation's wealth in the hands of the top 20 percent of our population. Bright people will attend the same universities and drink coffee together in the same cafes, marry one another and produce equally bright children who will continue to do the same. Because the brains and the bucks will be concentrated in this upper 20 percent, the larger 80 percent, with average or below average IQs, will have to settle for lesser-paying jobs or no jobs at all. This could prove expensive to the wealthy, who, through government programs, would have to provide support services—education, day care, medical care, police departments, prisons, and the like—for those unable to pay. Therefore, society as a whole would benefit if there existed fewer people with lower IQs. And, "the most efficient way to raise the IQ of a society is for smarter women to have higher birth rates than duller women."[41]

So, *The Bell Curve* authors zero in on unwed mothers. According to NLSY statistics, more than 25 percent of African American babies are born illegitimately; and the rate in low-income black communities is above 50 percent. White women with college degrees give birth to only 4 percent of white illegitimate babies, while women with a high school education or less account for 82 percent.

> Statistically, it is not good for children to be born either to a single mother or a married couple of low cognitive ability. But the greatest problems afflict children unlucky enough to be born to and reared by unmarried mothers who are below average in intelligence—about 20 percent of children currently being born. They tend to do badly, socially and economically. They tend to have low cognitive ability themselves. They suffer disproportionately from behavioral problems. They will be disproportionately represented in prisons.[42]

What is the prescription that Herrnstein and Murray draw from this description? Withdraw government funding for programs that support lower-class families such as services to aid unwed mothers and pre-school programs such as Head Start. Cut welfare. From *The Bell Curve* again: "The technically precise description of America's fertility policy is that it subsidizes births among poor women, who are also disproportionately at the low end of the intelligence distribution. We urge generally that these policies, represented by the extensive network of cash and services for low-income women who have babies, be ended."[43] If we stop subsidizing births to anyone, rich or poor, economic forces will lead to a proportionate rise in the percent of high-IQ babies born to married women in stable families who will require fewer government services and put fewer people behind bars. This is the gene myth in action.

One of the most egregious failures in sound judgment made by Herrn-stein and Murray is to reduce all problems to IQ genetics. That such a reductionism is taking place is not immediately obvious, because *The Bell Curve* authors frequently qualify their assertions by admitting that many factors contribute to the phenomena they analyze. They acknowledge that their statistics do not always show consistent correlations and that envi-ronmental factors such as divorce or welfare dependency contribute to undesirable social behavior. Yet they still proceed to assume that raising IQ without directly changing these other factors would correct our social problems. "Unfortunately," writes theologian Robert A. Pyne, "*The Bell Curve* treats intelligence as an independent factor even while the authors themselves decry that approach."[44]

The problem of unwed mothers simply cannot be reduced to IQ genet-ics. What is obviously decisive here is the social environment. The rate of illegitimate births remained virtually stable between 1920 and 1960 at about 5 percent of all births. With the 1960s it began to rise sharply, reaching 30 percent by 1990. Herrnstein and Murray admit that even if IQ is a factor it cannot be the only one. IQ must work in combination with other factors, because average IQ itself could not have dropped nearly enough in recent years to account for the explosive rise in illegiti-macy. What they fail to see is that this evidence points away from genes to environment. Critic Alan Ryan says they "evade the obvious implica-tion that their obsession with IQ is largely irrelevant." *The Bell Curve* authors argue that broken homes and the welfare system have interacted with IQ, making it more likely that a woman of low cognitive ability will have a baby out of wedlock. However, if broken homes and the welfare system are factors, then it is more likely that women of *any* degree of cog-nitive ability will have a baby out of wedlock. "The interesting question," says Ryan, "is not one of genetics but one of changes in the culture; it is not what happened to the intelligence of the mothers that needs explain-ing, but what happened in the early 1960s that so altered the incentives to have babies later rather than earlier and in wedlock rather than out. (It must mean something that divorce rates rose at the same speed during the same years.)"[45]

Head Start may deserve criticism, to be sure. But a proposal to aban-don such a program entirely because it supposedly serves those with lower cognitive ability does not follow from even *Bell Curve* premises. Failures in the Head Start program should be met by improvements in the pro-gram, by methods that work more effectively to bring economically dis-advantaged children into the mainstream. There is no genetically based warrant for taking resources away from Head Start and devoting virtually all our attention to those born to the cognitive elite. Ryan continues: "What one can certainly say is that the failure of Head Start to live up to its backers' most extravagant hopes is neither a knock-down argument for genetic determinism nor any sort of argument for abandoning the disadvantaged."[46]

To their credit, Herrnstein and Murray are aware of the inflammatory implications of putting together genes, intelligence, and race. The dominant view among molecular biologists and anthropologists is that scientifically speaking no such thing as race exists. The genetic differences between two individuals are greater than the average genetic differences between so-called racial groups, and the lines between these racial groups are blurred. Race is literally "skin deep," due only to migratory and other patterns of inbreeding over the last tens of thousands of years that set some groups on slightly different evolutionary paths. That racial differences are genetically inconsequential is the dominant scientific view. So, Herrnstein and Murray fly in the face of the dominant scientific view. The topic of genes-IQ-race in the late twentieth century is like the topic of sex in Victorian England, they complain. Back to *The Bell Curve*: "Publicly, there seems to be nothing to talk about. Privately, people are fascinated by it. As the gulf widens between public discussion and private opinion, confusion and error flourish. As it was true of sex then, so it is true of ethnic differences in intelligence now: Taboos breed not only ignorance but misinformation."[47] Such a defense is a form of self-congratulation for the courage to speak out. But just because they break a taboo does not mean that they have helped to dispel ignorance or misinformation.

It is *The Bell Curve* that spreads misinformation, say critics of the gene myth. The Joint Working Group on Ethical, Legal, and Social Implications (ELSI) of the Human Genome Project, jointly sponsored by the National Institutes of Health and the Department of Energy, has felt compelled to enter the controversy. ELSI ethicists argue that what we have learned from genetic research does not support the social policies advocated by Herrnstein and Murray.[48] Lori B. Andrews, who chaired the Joint Working Group panel, and colleague Dorothy Nelkin raise three points. First, even though Herrnstein and Murray invoke the authority of genetics to argue that cognitive ability is substantially heritable, the scientific jury is still out. Research in this field is still evolving. Geneticists are acutely aware of the enormous methodological difficulties in sorting out genetic components from environmental factors in the intricate interplay of nature and nurture in such a complex human trait as intelligence.

Second, *The Bell Curve* misrepresents lessons learned from genetics and from education. The assumption made by Herrnstein and Murray is that because cognitive ability is heritable that it is unchangeable, that it is not possible to alter it through remedial education. "This is neither an accurate message from genetics nor a necessary lesson from efforts at remedial education. Heritability estimates are relevant only for the specific environment in which they are measured. Change the environment, and the heritability of traits can change remarkably. . . . Height is both genetically determined and dependent on nutrition." A healthy and stimulating educational environment could have a positive affect on a child's functioning intelligence.

Third, the more scientists learn about genes the more complexity they discover. We are becoming increasingly aware of the degree of complexity in the interaction between genes and environment, and this inclines us to believe less in our ability to make predictions that are based solely on genetic information. "Simplistic claims about the inheritance of such a complex trait as cognitive ability are unjustifiable; moreover, as the history of eugenics shows, they are dangerous. . . . Genetic arguments cannot and should not be used to determine or inform social policy in the areas cited by Herrnstein and Murray."[49]

Back to Original Sin?

The gene myth with its puppet determinism forces us to ask a curious theological question: Might we be going back to the idea of original sin? Now, one might wonder what the mention of sin is doing in a book on genetic research and its impact on culture. Is sin not an anachronistic idea that we moderns discarded when we emerged from the Middle Ages? Our medieval forbearers listed seven deadly sins: pride, envy, anger, covetousness, sadness, gluttony, and lust.[50] Each refers to an inner disposition, a proclivity or tendency that leads toward violence either toward oneself or toward someone else. Despite the impatience the modern secular world has with premodern religious ideas such as sin, our world today is riddled with uncontrollable compulsions, aggression, violence, and resulting suffering. "Sin" is still the best word to use to get at this dimension of human experience.

But what about original sin? What about an inborn proclivity or tendency? Does this idea make sense today? After all, some within the Christian tradition have defined original sin in terms of an inherited propensity to do evil. Could the concept of genetic determinism inadvertently provide scientific support for inherited sin? This is not exactly the position I plan to defend; yet I believe it will be worth our while to follow the logic and see where it leads.

How would the concept of an inherited propensity toward sin sound to contemporary ears, especially secular or humanistic ears? Anachronistic?

Would it be unheard of for us in our modern liberal culture to think we have a natural tendency toward sin? Is it preposterous to agree with the Augsburg Confession of 1530 that all of us "who are born according to the course of nature are conceived and born in sin," or with John Calvin that original sin "seems to be a hereditary depravity and corruption of our nature"?[51] Was St. Paul deluded when he wrote in Romans 7:18–19 that "nothing good dwells within me, that is, in my flesh. I can will what is right, but I cannot do it. For I do not do the good I want, but the evil I do not want is what I do"? Is it impossible for us today to understand the human predicament to include a war within the self, a battle between our inherited propensities and the ethical ideal we project for ourselves, a struggle to overcome our natural origin by answering a call to a divine future?

One way the concept of original sin is understood is as inherited sin, as an inherited orientation toward evil. The Formula of Concord of 1580 says that we today inherit "an inborn wicked stamp, an interior uncleanness of the heart and evil desires and inclinations. By nature every one of us inherits from Adam a heart, sensation, and mind-set which, in its highest powers and the light of reason, is by nature diametrically opposed to God and his highest commands."[52] John Calvin describes newborn babies: "Even though the fruits of their iniquity have not yet come forth, they have the seed enclosed within them."[53] What could the idea of inherited evil desires mean in our era of genetic determinism?[54]

Will the debate over the crime gene provide material for theological consideration? Or, will the theology of inherited sin provide material for scientific consideration? Once these might have been thought to be anachronistic questions. But not now because, in the words of Dorothy Nelkin and Laurence Tancredi, "In the long debate over the relative influences of nature and nurture, the balance seems to have shifted to the biological extreme."[55] Here we ask: What is at stake in the Christian understanding of original sin now that nature seems to be shouldering more moral responsibility?

Manichaean, Pelagian, and Augustinian Alternatives

We can clarify just what is at stake in the idea of original sin if we look at three alternative directions taken by religious thinkers: the Manichaean, the Pelagian, and the Augustinian. The first alternative, the Manichaean, is to think that by nature we human beings are sinful. Sin is both naturally necessary and historically inevitable because we have been created as physical beings, and the physicality of the world leads necessarily and inevitably into the darkness of evil. Both nature and history are evil. And should we want to blame the divine, we could say that a god of darkness is culpable for creating nature the way it is and for making us the way we are. In rejecting this Manichaean view, classical Christian theologians repeatedly affirmed that the one God is good, and also that what God creates is good. What is natural is good, because God views it and says that it is "very good" (Genesis 1:31).

The second is the Pelagian alternative. Here human nature is essentially good. Sinful behavior is neither naturally necessary nor historically inevitable. People are born free. The human will is intact. Human history includes occasions in which individuals choose to do evil, of course; but we are free. If and when we choose to live by our God-given nature, then history will liberate itself from sinning. Nature is good, even if history contains blemishes. However, we need to ask Pelagians: is there sufficient reason to expect the human race to transform itself? Will this leopard change its spots? There is no warrant for thinking so. A realistic appraisal based upon the past human history indicates that sin appears inevitable for the future.

So we turn to the third alternative, the one that comes from Augustine. This fourth-century North African bishop affirmed that God is good and that nature is created good; yet he believed sin is something we can and should fully expect for the foreseeable future. Sin is naturally unnecessary yet historically inevitable.[56] Why? What we have since come to know as the doctrine of original sin is Augustine's answer. It goes like this. At an early point in history, God created the first human beings and called them good. We know our original parents as Adam and Eve; they committed the first sin. In Lamarckian fashion, this historical incident affected their biological nature so that the propensity for sinning became henceforth a congenital if not a heritable trait. Augustine likens this to a disease which we inherit congenitally and pass on to our children. Adam's original sin becomes for us virtually inherited sin, a predisposition that is not necessary by nature but for us is historically inevitable. In terms of human history, sin is the historical corruption of an originally good nature.

What led Augustine to this form of explanation was a fundamental commitment to the unity of the human race. This is a double unity, a unity of sin and a unity of salvation. He was taking with utmost seriousness passages in the writings of Paul such as 1 Corinthians 15:22: "As all die in Adam, so all will be made alive in Christ" (see also: Romans 5:12–16). The sins you and I commit today do not merely imitate Adam's sin. Somehow they participate and mutually penetrate and continue to propagate. Augustine was not primarily concerned with the specific propensities each of us has for lust or envy or whatever. That was secondary. Primary was this sense that all of us in the human race are in the same boat, the same sinful boat that is sailing away from consciousness of God and away from love one for another. This unity in sin is the correlate of our unity in salvation. Augustine writes that "we have derived from Adam, in whom all have sinned, not all our actual sins, but only original sin; whereas from Christ, in whom we are all justified, we obtain the remission not merely of that original sin, but of the rest of our sins also, which we have added."[57] For Augustine, the savior Jesus Christ is an individual human being, to be sure; yet he is much more. Christ is the prototypical human being, the eternal logos and the image of the divine—the true *imago dei*—under the conditions of humanity. The whole human race finds its definition, its identity, and its rescue from inherited sin through the forgiving and resurrecting power of Christ.

One of the major weaknesses in the Augustinian solution is that we in the modern world cannot split nature and history in quite the same way he did. Augustine could begin with the idea of a nature created and labeled by God "good," and Augustine could also identify a point in historical time—the time of Adam and Eve in the Garden of Eden—to locate the originating sin that corrupted the nature that we today inherit in our genes. There are three reasons that this is difficult to accept today. First, the current view informed by evolutionary history makes no room in its time scheme for an Eden story—that is, we cannot go back in time to find a point prior to which nature was benign and after which it had fallen. Second, evolutionary theory combined with such things as Big Bang theory in physical cosmology are leading scientists to view nature itself as history. Nature itself is neither fixed nor eternal, but subject to contingency and change. So, to bifurcate reality into two discreet realms, nature and history, no longer makes sense. The third reason is internal to theology. Despite what we have just said about the fall, Christian theologians including Augustine still assert that nature is good. Does this make sense? Is there a confusion at work here?

The Natural As Good Yet Corrupted

Today's mild confusion over the goodness of nature is due in part to previous unclarities in the history of Christian thought. Theologians of previous generations found it difficult to spell out clearly the connection between a good creation and original sin. The fundamental axiom is that creation as it comes from the hand of the divine creator is good. There have been two ways to say this, a pretemporal way and a nontemporal way. According to the pretemporal scheme, which we saw earlier as employed by Augustine, the world as a whole is originally created good.

Adam and Eve originally belonged to this good creation. Then they fell, and all of creation fell with them. We now live in a fallen world. It would be easy to solve the problem by simply leaving it this way: we could say that God created once at the beginning, but subsequent to the fall, creation is no longer good. This would easily explain our inherited predisposition to sin. Had we been born earlier, say back in the Garden of Eden, we would not have inherited this predisposition.

Yet, theologians for the most part have not been satisfied with following this easy path. Rather, they add a nontemporal approach. They insist that each person among us today is a special creation of God and, therefore, by nature good. Even after the fall, each of us along with the rest of nature is considered good in the theological sense. One reason for holding this is that God's creative work is not limited to a single act at the beginning. To say God creates is not to say that God called nature into being and then went off on vacation, leaving the world to run all by itself. This would be deism, not theism. So theistic Christians and Jewish theologians as well believe that God's creative work is ongoing. In addition to *creatio ex nihilo* once upon a time at the beginning, we have ongoing *creatio continua*. The audacious corollary to the axiom that creation is good is that creation is still good now even after the fall into sin.

The Formula of Concord of 1580 puts it this way: "Scripture testifies not only that God created human nature before the fall, but also after the fall human nature is God's creature and handiwork." It follows, therefore, that even after the fall we "cannot be identified unqualifiedly with sin itself, for in that case God would be the creator of sin."[58] To be sure, "God is not the creator, author, or cause of sin."[59] This commitment forces an equivocation in the way we use the word "nature." On the one hand, nature is that which God creates and, therefore, must be judged good. On the other hand, what we experience as sinners wrestling with our biological makeup are "evil desires and inclinations" put within us by nature. The single concept of nature has performed a double duty over the centuries, pointing to the goodness of God's creation while locating as inborn at least some of the temptations toward sin. No wonder confusion has occasionally appeared.

How do we sort this out? At least one observation we can make here is that we can deal with the concepts of a good creation plus original sin without necessarily placing them within a pretemporal scheme that begins with an original paradise followed by a historical fall. Nor are we stuck with the disease metaphor, according to which the germline DNA of Adam and Eve undergo a mutation that is then passed on to every generation thereafter. Both the history of paradise and the disease metaphor are attempts to frame a more basic picture. They are the clothing that dress an otherwise naked truth. What is that more basic truth? As we saw with Augustine's interpretation of St. Paul, it is the inescapable necessity to rely upon God's grace if we are to have a salvific relationship with God. The human race as a whole is united in its estrangement from God—one in Adam—and similarly united as the beneficiary of God's gracious action on our behalf—one in Christ. "Original sin" is the vocabulary we use to describe this state of estrangement from God which is the counterpoint to the state of grace created by God.

This has led theologians in the twentieth century to distinguish between sin as a state and individual sins as acts. Contemporary Roman Catholic theologian Roger Haight speaks of "the sinful condition of human existence as distinct from sinful acts."[60] In the previous generation neo-orthodox Protestant Paul Tillich described original sin as a state of estrangement. Particular sins represent the state of estrangement in action. "Sins are the expression of sin." It is not disobedience to a divine law which makes an act sinful; rather, it is the fact that it is an expression of our estrangement from God, combined with our separation from other people and even an expression of division within our own self. "Therefore, Paul calls everything sin which does not result from faith, from unity with God. . . . In faith and love, sin is conquered because estrangement is overcome by reunion."[61]

Even in the state of original sin, theologians have consistently affirmed the human race is capable of achieving great things. By nature we have the capacity to reason, to pursue science, and to be creative in the arts. We have the ingenuity to devise tools for building things. We have the moral sense to establish just and civil societies and to espouse high-minded philosophy. What we cannot by our own reason or strength do, however, is establish a relationship with God. The divine-human relationship is something for which God is responsible. To make this point is the task of the doctrine of original sin. To make this point it is not necessary to posit the historicity of the Garden of Eden or to employ the disease metaphor wherein we inherit a predisposition for sinning. The paradise story and the concept of inherited sin are the dressing for the otherwise naked proposition that God and God alone is responsible for establishing a divine-human relationship that is salvific.

Relationality and Solidarity in Sin

How can we describe the oneness the human race shares with Adam and Eve in conjunction with our individual propensity to sin? Theologian Marjorie Hewitt Suchocki has an answer. She uses the process metaphysics of Alfred North Whitehead as a springboard for developing a relational theology. Whitehead's organismic philosophy included the principle that everything is related—internally related—to everything else. For Suchocki this means that human subjectivity is internally related to both our DNA and to our environment—that is, we are who we are as an individual self only in relationship to others. Who we are as a person is internally related to our parents from whom we inherited our DNA and to the physical and social environments that continue to influence us. This relational metaphysics replaces Augustine's history of the fall as the conceptual frame within which she views original sin and sinful activity. This conceptual move is necessary, she believes, because the older conceptual framework is outdated. Yet the doctrine of original sin has much to say that is essential to understanding the human situation. Suchocki elaborates:

> The ancient church developed a notion of "original sin," clothed in mythic structures, that spoke to conditions set in force long before our individual births that nonetheless orient each of us toward sin. The notion carried conviction for more than a millennium of Christian history, but lost ground in the age of the Enlightenment. . . . Christian theology has been hard-pressed to retain the explanatory notion of original sin as the presupposition of the human condition. The issue is difficult: how does theology retain some explanation for the pervasiveness of sin, and yet at the same time avoid implicating God as creator of an imperfect creation? Human freedom alone cannot bear the burden.[62]

There is a dimension of the human experience of sin that is not reducible to free human acts. Somehow a dimension of human sin precedes our free will. We are born into it. We share it with the rest of the human race. A relational worldview can explicate this in a way that makes sense.

Suchocki's definition of sin is clear: sin takes the form of violence that contributes to the ill-being of any aspect of creation, to other people or other creatures or even to the planet Earth itself. Sin is rebellion against creation, and thereby, indirectly, rebellion against God. Sin has a trifold structure. First, we today have an inborn bent toward violence bequeathed to us by our evolutionary past. Although she does not rely upon sociobiology, she could at this point.

Second, interwoven relationality creates a solidarity of the human race. "We are individuals, but we are also participants in an organic whole much greater than ourselves, the human species. . . . Through the organic solidarity of the race, we are affected by the sins of others, and our own sins likewise have an effect on all others. . . . Evil anywhere is mediated everywhere through the relational structure of existence."[63]

Third, because of the temporal structures of intersubjectivity, our individual subjectivity inherits from our social environment the assumptions and values of the previous generation that filter our interpretation of the world. If those who came before us lived with assumptions and values bent in the direction of ill-being, then we are involved in support and perpetuation of that ill-being. But rather than place our predisposition to sin in our genetic inheritance, Suchocki follows Social Gospel theologian Walter Rauschenbusch and neo-orthodox theologian Reinhold Niebuhr in faulting our social inheritance. The values, norms, prejudices, hypocrisies, and structures of domination of the previous generation are passed along to us from our parents, and we reinforce them and pass them along to our children. Suchocki again: "The societies will be the bearers of sin to which the children will adhere even before they have the means of assent . . . before they have the means to exercise either consent or denial toward the corrupting sin."[64] This is relationality through time, intersubjectivity shared from one generation to the next. It is important that we see that social values as Suchocki understands them are not merely external, not merely environmental. We are internally related to these values because they become constitutive of who we are as a self. Suchocki discusses the self and sin:

> Since it is the individual *self*-consciousness that is so formed, it becomes constitutive of the self, and difficult to transcend. One's actions from this center of consciousness will then actualize the norms, perpetuating them relative to one's own position and perspective within the grid of the intersubjective society at large. By definition, the inherited norms cannot be questioned prior to their enactment: one is caught in sin without virtue of consent. Original sin simply creates sinners.[65]

Original sin comes first, then sinners who sin. This leads Suchocki to exclaim, "This means, somewhat paradoxically, that one can be a sinner innocently."[66]

If we think of the two-part determinists who have been fighting the gene myth, it appears that Suchocki would be a kindred spirit. She affirms that we inherit a predisposition to sin, but it is nurture and not nature that bequeaths to us this inheritance. Is Suchocki a determinist? Does Suchocki believe in human freedom? Yes, to both. Determinism and freedom are not incompatible for Suchocki. We are clearly determined by our DNA and by our social relationships. Yet there is more. As persons we are able to respond. We are "response-able." In fact, it is our responses to the determining agents within and without that create the self. Through response the growing self transcends its determining conditions. "Freedom is always conditioned . . . while freedom is necessarily conditioned by one's past and one's context, freedom cannot be reduced to those causes . . . "response-ability" is at the core of every moment of our lives. . . . This freedom is an ability to question the givenness of oneself and one's world."[67]

Suchocki has an antidote for sin: forgiveness. "Forgiveness as the act of willing the well-being of the other is a direct intervention that has the power to break the cycle of violence."[68] Forgiveness releases the violation and consigns it to the past. This release into time opens the door to transformation for the future.

I believe that this consignment of sin and its evil consequences to past time also opens the door a crack for the theologian to glimpse a way of putting together the concept of creation's goodness along with the concept of original sin. Note that we have ruled out limiting God's creative work to a single event at the beginning. We have added to it the idea of continuing creation, noting how classical theology has described the birth of each one of us as God's bringing something good into the world. With relationality we get holism. Add time and we get temporal holism. Perhaps we could say that God's creation is not done yet; it is not yet the whole that it will someday be. Perhaps we could say that transformation is part of the divine creative process, so that the creation will not be complete until it is transformed, until it is redeemed. Is this not an implication of the eschatological promise of the new creation?

I would like to offer a theological hypothesis: we can think of the whole of creation as inclusive of its time. With this we will have a way of conceiving of the whole of fallen humanity united in Adam simultaneously with the whole of redeemed humanity united in the new Adam, Jesus Christ. The present era or aeon, replete with human achievements combined with an inherited predisposition to sin—whether inherited genetically or socially or both—belongs to the era or aeon of Adam and includes all of us.

If we factor in time and transformation and the vision of paradise before the fall, and if we rule out locating paradise in the past, then we might look to the future. Perhaps we could think of a transformed future wherein original sin would belong only to the past, the dead past. This transformed future is referred to in the Bible with symbols such as the kingdom of God or the new creation. It is the realm of freedom promised us by God.

Now, we might add a caveat: either there is a God who can make such a promise or there isn't. And if there is such a God, then either this promise will come true or it won't. Be that as it may, the promise belongs at the heart of Christian theology. Its confirmation awaits the future. Theologically speaking, of course, future confirmation has arrived ahead of time at Easter, at the time when God raised Jesus from the dead.

Easter distinguishes the aeons of Adam and Christ, of the dead and of the resurrected. "As all die in Adam, so all will be made alive in Christ" (1 Corinthians 15:22). The promise for the new aeon is that resurrection will become part of what it means to be human.

The aeon of Jesus Christ is the future new creation. This future new creation embodies unambiguously the good. Yet that future has appeared ahead of time. The new aeon has become incarnate in the old aeon. Jesus Christ is the prolepsis—the incarnate anticipation—of God's final future. In his Easter resurrection, he anticipates the new creation. Easter is the new creation appearing ahead of time in this person. And, as the Holy Spirit makes the resurrected Christ present to us in faith, the qualities of the future new creation become present to us now within the old creation. Recalling from the first chapter Luther's description of faith wherein Christ is actually present, we can see how we can be at one and the same time both fallen and redeemed. Simultaneously, we are both sinners and new creatures.

Conclusion

We have three ways to understand the concept of original sin: (1) as a categorical description of our state of estrangement from God and our need individually and as a single human race for God's transforming grace; (2) genetically as an inborn predisposition or orientation toward expressing oneself in ways that do evil; and (3) environmentally as sin that originates outside of ourself but incorporates us in its work of social evil. Of the three, the first is essential to Christian theology. The second two might even be thought of as ways to describe how the first one could be understood.

Another way to describe the state of estrangement over against God's reuniting grace is temporally through the concept of the two aeons or two eras. The present aeon—the aeon that may have begun with the cosmic Big Bang twenty billion years ago and has undergone hundreds of millions of years worth of evolutionary history—is the aeon of Adam. Even if the gene myth were to be proven valid, it would apply to Adam's aeon. We could say with the evolutionary psychologists that nature's history has placed us here with selfish genes that have destined human society to constant clan warfare in the competitive race to see whose genes will win. Or, we could say with molecular biologists and behavioral geneticists that we have specific genes that predispose us to specific forms of behavior that lead to violence in our relationships. Genetic determinism might be able to explain our past, and in doing so it would provide a way of conceiving of both our unity as a human race and our inherited propensity to engage in some forms of evil.

If, however, the critics of the gene myth prove to have the better case, then we would find ourselves emphasizing a two-factor rather than a single-factor determinism. Environmental influences would be taken much more seriously than they are by the gene myth. Marjorie Hewitt Suchocki's relational understanding of solidarity fits well here. It presupposes the oneness of the human race while providing a social version of inheritance that explains how original sin is passed from one generation to another. Whether biologically transmitted or socially transmitted, when we wake up to consciousness we find that we have been born individually or in groups with the inherited propensity for violence. We humans are all in this together.

We live in the present creation, Adam's aeon. As such, it is ambiguous. Our evolutionary history has placed the present generation in a situation of great opportunity. If we draw upon our natural endowments, such as our ability to reason along with our ability to envision justice and beauty in human society, we as protheans could determine a future that would bring better health and increased well-being to the whole of the human race. Or, if we surrender to certain genetic and social determinants we have inherited, we instead will enlist our newly acquired scientific and technological powers to serve the interests of class privilege and racial prejudice. We clearly have a predisposition for both.

< FOUR >

The So-called "Gay Gene" and Scientized Morality

Genetics provides *a basis for grace within the structure of life itself.*

—Lindon Eaves and Lora Gross

In the summer of 1993 a scientific bomb exploded that is still causing considerable ethical fallout. Dean H. Hamer and his research team at the National Cancer Institute announced that they had discovered evidence that male homosexuality—at least some male homosexuality—is genetic. Hamer and his team constructed family trees in instances where two or more brothers are gay and combined laboratory testing of homosexual DNA. They located a region near the end of the long arm of the X chromosome that they believe contains a gene influencing sexual orientation. Because men receive an X chromosome from their mother and a Y from their father (women receive two X's, one from each parent), the possible gay gene is inherited maternally. Mothers can pass on the gay gene without themselves or their daughters being homosexual. A parallel study of lesbian genetics is as yet incomplete; and the present study of gay men will certainly require replication and confirmation. Scientists do not yet have indisputable proof. Nevertheless, Hamer was ready to write the article making the dramatic announcement, "We have now produced evidence that one form of male homosexuality is preferentially transmitted through the maternal side and is genetically linked to chromosome region Xq28."[1]

The press exploded with follow-up articles. *Time* projected an ethical and political forecast: "If homosexuals are deemed to have a foreordained nature, many of the arguments now used to block equal rights would lose force." *Time* cites a gay attorney: "I can't imagine rational people, presented with the evidence that homosexuality is biological and not a choice, would continue to discriminate."[2] The logic seems to be this: if male homosexuality is genetically inherited, then those with this inheritance can claim rights based on this inheritance. If sexual orientation is genetically determined and not a choice, then gay men can claim the ethical high ground. Yet, I ask: Is this the immediate conclusion to which we should jump? Is this the only direction we could go? Are there other possible directions?

I would suggest here that if eventually we accept as fact that male homosexuality is genetically inherited, then the ethical logic that follows could go a number of different directions. *The scientific fact does not itself determine the direction of the ethical interpretation of that fact.* To demonstrate this I would like to begin with a basic question: Does the genetic predisposition toward homosexuality make the bearer of that gene predetermined or responsible, innocent or guilty? Two answers are logically possible. On the one hand, a homosexual man could claim that because he inherited the gay gene and did not choose a gay orientation by his own free will, he is morally innocent. His status as a gay man should not count as sinful nor should he be victimized by social discrimination. He might further claim that if homosexuality is his by nature, then he should have the moral and the legal right to express his inborn nature.

On the other hand, we could take the opposite road and identify the gay gene with the predisposition to engage in socially unacceptable behavior, what some might call sin. Society could claim that the body inherited by each of us belongs to who we are—who I am as a self is determined at least in part by what my parents bequeathed me—and that an inherited predisposition to homosexual behavior is just like other innate predispositions such as lust or greed or similar forms of concupiscence which are shared with the human race generally. All this constitutes the state of original sin into which we are born. Society can further claim that each of us cannot escape responsibility for our own body, for our own self, and still further that through the exercise of the human will, we ought to choose the good over the bad and guide our physical predispositions toward a virtuous and socially constructive lifestyle.[3] Signposts point in both ethical directions.

Those among us wishing to liberalize social attitudes regarding homosexual activity may object that the very use of the word "sin" is an anachronism. The concept of sin seems harsh, judgmental. The theological use of sin, it is feared, parallels the secular use of the concept of normality. If heterosexual activity leading to procreation is assumed to be normal, then homosexual activity appears abnormal or deviant. When fighting for gay rights it is important to presume that gay and lesbian activity is not sinful, not abnormal, and therefore should be socially acceptable. Short of social acceptability, at least homosexual activity, like all sexual expression, could be thought to be one's own private matter. As long as no one is victimized, our ethics and legal practice should protect private homosexual expression.

But this position begs the question. It assumes that homosexual expression is normal and, therefore, not sinful. A prior question is: On what grounds can we consider it normal or not sinful? It is this underlying question that animates the public discussion over the gay gene. The sometimes unarticulated yet pervasive hypothesis is this: If science can show that homosexual orientation is genetically determined—that is, normal—then can we theologically and ethically treat homosexuality as natural and thus not sinful? Our task here will be to track the various paths that ethical and theological thinking might follow if credibility can be attributed to the idea of genetic predisposition to homosexual orientation.

Is There Really a Gay Gene?

Can we say with scientific confidence that there *does* exist a biological basis for homosexual orientation? No, not yet. But the idea seems plausible, and the time for theological and ethical reflection on this idea is now.

The discovery of Xq28 in 1993 fits within a more inclusive history of the search for a biological explanation for homosexual orientation. Might we find the explanation in hormones? Endocrinologists have advanced the testosterone—or lack thereof—hypothesis. The hormone testosterone distinguishes girl from boy fetuses and influences the development of internal genitalia, external genitalia, and the brain. Working on the assumption that gay men lack masculinity, some physicians have tried to "cure" their patients with testicular grafts or even injections of testosterone. The result was not a sexual reorientation toward women. Rather, the desire for wanting sex with other men only increased. It seems clear that hormones act on the brain to stimulate desire, but just how they act on the brain is as yet unknown.

That scientific attention would turn to the relation of the brain to sexuality could be expected. In the summer of 1991 neurobiologist Simon LeVay, then with the Salk Institute for Biological Studies in La Jolla, California, claimed to have found a difference in brain structure between gay and straight men. LeVay focused on a part of the hypothalamus, a small area near the pituitary gland at the base of the brain. It had been known previously that in both rats and humans the hypothalamus is smaller in females than in males. Hormones acting on the hypothalamus in a female rat prompt her to perform lordosis—raising her rear to the male to facilitate intercourse. When testosterone is injected into the medial preoptic area of the hypothalamus of a castrated male rat, he will still try to mount the female in lordosis. This animal behavior suggested to LeVay that the medial preoptic area could be crucial to male sexual behavior.

So LeVay examined tissue taken from forty-one routine autopsies of gay men, straight men, and straight women. He cut the hypothalamus into paper thin slices and examined them under a microscope. The microscope examination of these cells was done as a blind study, using identification codes, to avoid prejudicing the procedure. LeVay focused his attention on the third interstitial nucleus of the anterior hypothalamus known as INAH–3. He found that on average INAH–3 in heterosexual men is two or three times greater in volume than in homosexual men and in women. This could mean that the cell group's smaller volume is responsible for the homosexual orientation of the males investigated. Critics have argued that the twofold difference in cell size is too small to matter; but LeVay insists that statistical tests reveal that the difference is significant. LeVay offered the modest conclusion that this finding suggests that sexual orientation has a biological substrate—that is, homosexuality may be innate.[4] LeVay goes on to speculate that if sexual orientation is somehow a function of the genetically controlled sexual differentiation that occurs in the early fetus, then it must somehow be influenced by genes.

By fall 1991 the search for the gay gene had begun by molecular biologist Dean Hamer and his team of researchers at the National Cancer Institute on the campus of the National Institutes of Health in Bethesda, Maryland. Hamer and colleagues conducted interviews and took blood tests from forty pairs of homosexual brothers to see if they shared a trait: specifically, to see if they shared a specific segment of the X chromosome. Rather than a shotgun review of DNA sequences on all twenty-three chromosomes, preliminary family studies had led Hamer to target the X chromosome, a chromosome on which previous molecular biologists had already identified a number of markers. Because any two brothers have a 50 percent chance of inheriting the same X chromosome from their mother, the commonality would have to occur in more than half the sets of brothers studied. The first of the Hamer team findings was that thirty-three of the forty pairs of gay brothers were concordant—that is, they shared a segment of DNA on the q arm of the X chromosome, designated on gene maps as Xq28. Thirty-three out of forty is 83 percent, which is considerably more than the 50 percent chance possibility. Other statistics showed a better than 99 percent probability that the observed linkage between the genetic marker and homosexual orientation was genuine. The second finding was that sexual orientation was not strongly linked to any other region on the X chromosome, only Xq28. The third finding was that the gay men have more maternal than paternal male relatives who also are gay, corroborating that the gay gene is passed by the mother.[5]

Hamer thought he had discovered something significant. "Now by studying DNA we'd found a connection between homosexuality and the chromosome itself," he writes. "We didn't know what the gene did or how it worked—nor, for that matter, exactly which gene it was—but we had strong evidence that it existed."[6] This first discovery still requires future replication and confirmation. "We hadn't actually isolated the gay gene, only detected its presence, and the results would need to be replicated and confirmed on an independent sample."[7] As of this writing, no independent laboratory has replicated or confirmed the work either of LeVay or the Hamer team.

To answer the question "Who is gay?" the Hamer team had relied on self-identification. Using the Kinsey scale, wherein 0 identifies a heterosexual with no homosexual tendencies and 6 identifies a strict homosexual with varying degrees of bisexuality (or ambisexuality) in between, the gay men Hamer interviewed clearly located themselves with integers toward the 6 end of the continuum.[8] What Hamer believes he and his colleagues found was a correlation between self-appropriated homosexual identity and a shared marker at Xq28. Significant here is that homosexuality is not defined biologically; rather, a self-appropriated social definition is correlated to a biological finding. No attempt has been made as yet to assert that the gene at Xq28 *causes* what these brothers experience as homosexual orientation. The hypothesis yet to be explored is that, should we find a gene at this location, it would predispose or influence a person's sexual orientation.

The fact that this genetic marker is found on the X rather than the Y chromosome leads to other possibly significant lines of thought. For centuries it has been assumed that homosexual men are womanish and lesbian women are mannish. Cultural images of the sissy boy and the tomboy girl reinforce the commonly held view that homosexual people are a kind of third sex, existing somewhere in the middle of a continuum between the strictly male and strictly female genders. Even Simon LeVay may have presupposed this by looking for what he found, namely, that the INAH–3 region of the hypothalamus would be approximately the same size for women regardless of sexual orientation and for homosexual men but larger for heterosexual men. In other words, gay men are like women in some biological respects. But, we might ask, will this be confirmed at the molecular level? The Hamer study leads us away from this likelihood.

Relevant here is the role that the Y chromosome plays in gender differentiation and one of its genes known as the testis determining factor, or TDF. The Y chromosome appears only in men, not women. At conception all of us are, for all practical purposes, female. At a later stage of fetal development, the TDF on the Y chromosome switches on two things. It first turns on the gene for a protein known as the Müllerian-inhibiting hormone, which prevents the formation of the female genital tract. Then, second, it turns on another set of cells to produce testosterone; and it is testosterone that stimulates the development of male genitalia and the masculinization of the whole body. Two switches get tripped by TDF, one to turn off the female pathway and another to turn on the male pathway. The result is the difference between boys and girls.

But, does the TDF also determine sexual orientation? Is sexual orientation somehow linked to TDF? If a developing male fetus gets a full blast of TDF will he turn out heterosexual, but in certain cases where the TDF functions only partially is the fetus destined to be a male homosexual? Conversely, will female fetuses with no TDF influence turn out heterosexual girls, but in the event that some TDF influence creeps in will they become lesbian? Is sexual orientation linked to the Y chromosome and to the full or partial functioning of TDF? This is a fair set of questions. The research of Dean Hamer and his colleague, Nan Hu, lead to negative answers to this line of inquiry.[9]

Cytogeneticist Nan Hu conducted related experiments and found that lesbian women, just like heterosexual women, are normally XX; and gay men, just like heterosexual men are XY. "These findings were confirmed by directly analyzing the TDF gene coding sequences by the polymerase chain reaction (PCR) method. As expected, the TDF sequences were present in all the men and absent in all the women, regardless of their sexual orientation."[10] Even though gender is determined by TDF on the Y chromosome, sexual orientation appears not to be. If there exists a genetic influence on homosexual or even heterosexual orientation, it must lie elsewhere. Could it be in part or entirely at Xq28? Could it be Xq28 that influences the hypothalamus?[11] If so, might we be able to add the following: At the molecular level, does there seem to be no room for a third gender, no room for an intersex status that may be causally related to sexual orientation? Hamer says, "The idea that gay men are like women, and lesbians are like men, is one of the oldest, and seemingly most obvious, concepts in sexology. The only problem is: it's not true."[12]

Hamer and his team have been subject to numerous criticisms. Many of the criticisms have to do with controls in the study: they should have examined the straight male siblings of the gay brothers, or they should have collected DNA from more than just fifteen of the mothers. There is a string of similar complaints.[13] Another criticism is that Xq28 may connect not to homosexual orientation but rather to a more general temperamental inclination toward a whole spate of behaviors of which homosexuality is only one. One wag has said facetiously that the alleged gay gene may correlate only with a willingness to discuss one's sex life in an interview with a geneticist.[14] At this writing an NIH investigative panel is formally reviewing the methods employed by the Hamer study. Until Hamer's conclusions are replicated or the credibility of his study undermined, we should work with the as yet unconfirmed hypothesis that a genetic predisposition for some forms of homosexuality exists in some men.

We find a category of criticisms worth mentioning that come from ideological warfare, specifically from gay social constructionists. "All attempts to find the origins of homosexuality in genes have failed," writes Rainer Herrn.[15] This negative judgment is based on the notion that all biological explanations are reductionistic while homosexuality is by definition psychological, social, cultural, and political. So it follows that "biological explanations, however they are worded, are insufficient."[16]

The assumption of social constructionists is that concepts such as homosexuality or heterosexuality are strictly social constructions imposed by culture. They are not biological categories, nor can they be adequately attended to by the reductionist methods employed in the natural sciences. They object to the vocabulary of sexual *orientation*, because it smacks of biological determinism; whereas the vocabulary of sexual *preference* keeps sexual expression in the realm of choice. If I am *oriented* genetically, then I'm not free. If I *prefer* homosexual rather than heterosexual expression, then I am presumed to be free. Social constructionists emphasize that gay men and lesbian women vary considerably in their sexual expression, sometimes changing at different stages in their lives. Sexual preferences change over time. They are not mere products of biological determinants.

Much of the social constructionist argument hinges on the complexities and nuances that inhere in the phenomenon we know as human sexuality. So much more than the mere biological is involved. The problem with biological research on hormones, brains, or genes is that it is reductionistic—that is, it reduces the phenomenon to what happens to the body. Omitted here is the intensity of one's personal history, the social setting, cultural influences, and the particular situation in which sexuality comes to expression. John P. DeCecco and David Allen Parker put it this way: "Certainly, as biological organisms, any and all of our behaviors must have biological correlates, but that does not mean that those correlates *determine* our behavior. . . . Sexuality can be reduced to neither a purely biological state nor a purely psychological one. Any plausible explanation of sexual expression would have to include all of its components."[17] Biological influences are not totally dismissed by this point of view. What is emphasized is that the way we feel and the way we interpret our actions is socially constructed.

Dean Hamer responds flippantly that "the social constructionist theory is not likely to be disproved any time soon, since its content is too amorphous to ever be tested."[18] Yet, this is too glib. In fact, Hamer agrees that human sexual expression is a wondrous and complex phenomenon that cannot be reduced to biology. When contrasting human from animal sexuality, Hamer humorously points out that "Pigs don't date, ducks don't frequent stripper bars, and horses don't get married. Anyway, since when are animals good role models? The praying mantis devours her mate while they are still copulating. Male dogs will mount their daughters. Animals don't speak, write love songs, build churches, or do a lot of other things that we consider worthwhile."[19] If human sexuality cannot be reduced to its animal prototype, it certainly cannot be reduced to its genetic predisposition.

As we conclude this portion of our discussion, we must acknowledge that the science is in doubt. Because so many human characteristics begin with our biological substrate, and this biological substrate is heavily influenced by our genes—or at least our individual genomes as they interact with our particular environment—it only stands to reason that we are likely to find a genetic predisposition to at least some component to what we know as homosexual orientation. If Dean Hamer's research team does not have it yet, we are hiding our heads in the sand if we are not looking for another research project to discover one. No amount of social constructionist ideology can wish away our DNA. Our task now is to engage in hypothetical speculation—that is, we need to ask "What if?" We need to ask about the ethical and theological implications of the possibility that such a genetic predisposition will be eventually confirmed.

We turn now to speculative reflection by drawing a map. We want to find out just where scientific knowledge regarding a gene for homosexuality might lead us. It is by no means a foregone conclusion that such scientific knowledge will lead to a peaceful end to the ethical and social turmoil we are experiencing over homosexuality. Nor will it lead automatically to increasing social acceptability and protection against discrimination for the gay and lesbian persons among us. It leads in different directions. To track the different directions is our next task.

My Genes Make Me Innocent

Two ethical roads diverge from the scientific woods, one leads to the claim of innocence and the other to the idea of inherited sin. Let us follow the road toward biological innocence first, looking algorithmically at some of its forks and intersecting side streets.[20]

Selecting the right term for this discussion was difficult for me. I have elected the term "innocence" meaning guiltlessness or freedom from moral wrong. This vocabulary may appear morally loaded to gay men or lesbian women who feel that their sexual orientation is innate to their being, who do not feel this state of being is blameworthy. It just is. Blame or innocence should apply only to how we act or what we do, not to our state of being. Yet, this is just what is at issue. The gene myth suggests that what we do is predetermined by our genes. The public debate is over morality or immorality of homosexual activity. The question is: Can science resolve the debate by declaring homosexual activity moral on the grounds that it is due to genetic determination? It is the logic of this debate I wish to track.

The logic of *biological innocence* might go like this: if we can establish scientifically that the basis for sexual orientation lies in the genes we inherit, then this puts sexual orientation into a category comparable to that of height or eye color, or nose size or similar bodily features which in themselves are natural and have no moral status. In fact, to discriminate on the basis of height, eye color, or nose size would in our society be considered immoral. It would follow that to discriminate against someone with the gay gene would be immoral. This would be a naturalist argument: if it has been determined by nature, then we have no choice. Or in its stronger version: if it's natural, then it's good.

A variant would be progay essentialism. An essentialist position holds that a trait is deep-rooted, fixed, and an intrinsic feature of a person. What we might call "biological essentialism" would require that this trait be rooted in nature, that it be genetic or otherwise biological. What we might call "personal essentialism" would tie this trait to the person, whether it originated in the biological substrate or some other dimension of the person's being. The biological innocence road is being followed by many progay biological essentialists who contend that homosexual orientation is fixed, immutable, and definitional to the individual person.[21] This renders their homosexual orientation innately good, or at least immune to outside interference.

Undoubtedly this will be a strong argument raised by gay rights advocates in those cases where their opponents presume that the homosexual lifestyle is merely a matter of choice. People we might place in the antigay constructivist camp hold that homosexual practices are immoral; they presume that homosexuality is voluntary, that individuals are enticed into homosexual behavior by seducers or by a gay subculture. Former Vice President Dan Qualye, for example, announced during the 1992 presidential campaign, "My viewpoint is that it's more of a choice than a biological situation. . . . I think it is a wrong choice."[22] The Family Research Institute, for another example, advances the "recruitment" hypothesis, saying that gays and lesbians "recruit" heterosexuals into a lifestyle of homosexuality.[23] Moral responsibility, according to this point of view, is found solely at the level of nurture and not nature—that is, it lies on the shoulders of recruiters and the individuals making the choice either for or against this allegedly perverted activity. However, if we can claim that homosexuality is not voluntary—that is, our predisposition toward homosexual behavior is not a matter of free will—then the logic changes. If it is biologically inherited, it is not free. We are not responsible for how we are born. We cannot be judged responsible for what is or is not in our DNA. Freedom, choice, and responsibility all go together, we presume; and we are denied freedom, choice, and responsibility in matters of genetic inheritance.

The road of biological innocence now forks. As suggested above, one branch would treat the gay gene as natural, essential, and at least benign if not outrightly good. The other branch would treat the gay gene as socially disruptive and hence undesirable, perhaps even as a genetic defect. Now, let us follow the branch which presumes that homosexual orientation leads to immoral behavior, placing the gay gene in a category with other undesirable genes—that is, genes that are defective or cause disease. We will call this the *biological innocence with a gay gene defect* branch. Later, we will return to our fork and follow the other path.

Would breast cancer count as a parallel case here? The gene or configuration of genes disposing millions of women to breast cancer known as BRCA–1 was recently located on chromosome 17. Scientists such as Mary-Claire King at the University of Washington are working to try to find the switch that turns the gene "on" or "off." The next task will be to invent a medicine to keep the gene turned off. The hoped for result will be protection against one of most terrifying killers of our time, breast cancer. This is the stuff out of which Nobel prizes are made. If this scientific goal is achieved, all of society will give praise.

Should we place the sought-after gay gene on the X chromosome in this undesirable category? Should we look for the DNA switch and try to turn it off? Should we greet genetic science with praise for putting us on the brink of finding a medical way to cure our society of homosexuality, one of its supposedly immoral diseases? Even though the individuals who carry the gay gene can be judged innocent just as women suffering from breast cancer are innocent, we still need to ask: Should we try to help both the victims of the gay gene as well as the larger society through medically eliminating the predisposition to homosexual behavior?

If the answer here is yes, then we will have two ways of curing our society of disruption due to homosexuality. The first would be to employ genetic medicine as therapy for existing gay men (and perhaps eventually lesbian women). We could ask the scientists to try to invent a medicine that would turn off the gay gene, presumably to transform existing gay persons into straight persons.

The second way would be to employ genetic engineering to prevent future gay people from coming into existence, from being born. Prospective parents could ask genetic engineers to provide methods of in vitro fertilization (IVF) that would eliminate the gay gene at conception. Or they could ask that the already conceived fetus be tested for the gay gene and, if found, be aborted. Biological essentialism, thought initially to be progay, can easily become antigay.

This branch of the innocent road now leads directly to an intersecting side street, namely, selective abortion. We can safely forecast that as scientists progress in mapping the human genome, the practice of abortion-on-demand will increase. As we increase our ability to identify undesirable genes such as those that cause disease, it will become almost normal to test fetuses and abort those with undesirable genes. The practice is already widespread for unborn children with Down Syndrome. Each newly discovered disease-causing gene will stimulate a new testing procedure and add one more reason to abort. We can expect abortion to become more prevalent as more and more of the suspected five thousand genetic diseases are located on the DNA map. We will presume that the unborn fetus is innocent, of course. It is not guilty for bearing the gene that causes it to be aborted.

Now, what will happen if we include the gay gene in this category? Will we find parents aborting fetuses out of fear of raising a homosexual child? Abortion practices in China and India already demonstrate that unborn females are aborted at a higher rate than males. Could we expect that unborn homosexuals will suffer a similar fate? Should they suffer a similar fate? Should we see abortion as a means for cleansing our society of a gay population?

John Tolin, writer of the play *Twilight of the Golds* wherein a woman considers aborting a fetus after a test showing the child may be gay, fears the logic: "Well, we've found the cause of this terrible thing and now we should try to cure it."[24] What we are talking about here could be called "negative eugenics" because it leads to a program for systematically eliminating from the human race a certain class of genetically determined individuals.

In sum, to follow the road of the innocent gay person with the undesirable gay gene may not provide the moral muscle necessary to prevent discrimination. It may provide relief from some forms of discrimination for gay persons living today but no protection to gays yet unborn. Conversely and ironically, those who would like to cure society of homosexual disruption may find that the therapeutic method—namely, abortion—is just as morally repugnant as the disease.

Let us return to the fork in the innocent road. We have seen where the *biological innocence with the gay gene defect* branch takes us. Now, where would the *biological innocence with a morally benign gay gene* branch lead? Not everyone in our society agrees that homosexual orientation or even homosexual behavior is immoral. For gay advocates, homosexual behavior is morally neutral or even morally desirable. Some object to applying the label "normal" to heterosexuals and "abnormal" or "defective" to homosexuals. So, the advantage in finding a genetic source for homosexual orientation could be to normalize. If it's natural, it's normal. According to this position, the gay man would be doubly innocent: innocent first because his sexual orientation is determined and not freely chosen and, second, because his homosexual behavior is just as morally acceptable as that of a heterosexual.[25] Now, where would this lead? Going this direction we would presume that homosexuality is like eye color—that is, quite ordinary, quite natural, one more example of genetic expression among others.[26] Or, alternatively, we might believe that the gay gene is an actual benefit. After all, some of us prefer blue eyes over brown, while others prefer brown over blue. Perhaps homosexuality could fit this category. We could switch it on or off to suit us, and some parents could abort gay fetuses while others abort straight fetuses. Homosexuals and heterosexuals would have equal rights to express their sexual preferences regarding progeny. The desired outcome here would be that no moral code would determine preference in advance, and no systematic discrimination program would be ethically licit to eliminate homosexuality.

But this approach also lacks the moral muscle to withstand the pressure to discriminate against homosexuality, because it places moral responsibility in free choice; and in our bourgeois society this means placing moral responsibility in the free market. It grounds selective birthing in subjective taste. Fears of a "brave new world" have already been aroused over the prospects of "designer genes" wherein the physical features of future generations could be the result of cultural fads and parental whims. Blue eyes cannot claim any intrinsic moral superiority over brown, nor the reverse. So, in principle, our children's eye color will be determined by the personal tastes of those giving engineered birth. To place the gay gene in this category would make it similarly subject to cultural values and parental choice. And should our cultural values and parental choice be influenced heavily by homophobia, then it would be quite predictable what would follow. All that we said above regarding IVF and abortion would apply again here. Because in our society children as yet unborn have no rights, one cannot expect that this branch of the biologically innocent road would detour discrimination.

Anticipating the possibility of selective abortion that might amount to gay genocide, Dean Hamer speaks out. "I think that discriminating against people based on their genetic makeup is wrong. . . . The 'wrong' genes should never be used as a basis for terminating a pregnancy."[27] Hamer opposes the idea of developing prenatal tests for homosexual orientation. If he eventually isolates the gene, he plans to patent this knowledge as intellectual property and use his patent rights to prevent the development of such tests and such selective abortion. "For all his good intentions, though, Hamer should know better," comments Daniel Kevles. "Except in extraordinary cases, the life of a patent is now only seventeen years." Eventually the knowledge will be available for this purpose. And because of the way abortion is protected in the United States—the decision is strictly the free choice of the mother—one cannot legally introduce criteria by which some choices are considered right and others wrong. The reasons women give for choosing to abort "may be morally endorsable, or they may be morally deplorable. If the right to choice is to have meaning, it cannot be diluted" by our desire to protect unborn gay fetuses from discrimination.[28] In sum, a naturalist argument that declares a person carrying the gay gene to be morally innocent will not in itself provide total protection against discrimination.

Let us go back to the map. The road that carries us toward the innocent gay person with the benign gay gene also intersects with an important side street that to this point does not appear on any moral roadmaps I have seen. This side street would ask about the genetic status of two other relevant human phenomena, namely, heterosexual orientation and gay bashing. If a gay gene exists, might we ask our scientists to return to the laboratory to find a genetic cause for heterosexual attraction? Is there a gene or complex of genes that attracts women to men and men to women? On which chromosome does it lie? Can it be turned on or off? If we find the answers to these questions, can we expect a whole new market in aphrodisiacs?

Similarly, might there be a gene that disposes some persons to homophobia and thereby to gay bashing? Is there a road with a sign: *Biological innocence with a gene for homophobia?* Homophobia—the predisposition to gay-bashing behavior—is so widespread in the human race that it is reasonable to think it could be genetic. In our society, of course, gay bashing is deemed immoral and illegal. Yet, we have to ask: Should we assume that homosexuality is genetic while homophobia is cultural? Should we assume homosexuality is due to nature while homophobia is due to nurture? This would be inconsistent. So, in honesty we need to ask further: Should we permit or encourage scientists to send a comparable search party into our DNA looking for the homophobia gene? And, if a gene for homophobia is found, should we declare the carrier of that gene innocent? Might we ask individuals with the homophobia gene to claim the same innocence and to express themselves with the same moral privileges we would grant those with the gay gene? Would the arguments that the inherited propensity toward homosexual orientation repeat themselves in the case of homophobia?

If we find ourselves convinced that homophobia belongs to our nature as much as homosexuality does, would we alternatively want to argue that one is benign but the other defective? Would we want to say that those possessing the gay gene are innocent, but those possessing the homophobia gene are sinful? Is there moral warrant for asking homophobiacs either to control their natural propensity or to take radical steps to eliminate their homophobic orientation? Are there any grounds for declaring homosexuals morally innocent by reason of genetic inheritance yet hold homophobiacs guilty for the same reason?

My Genes Make Me Guilty

This brings us back to the original point where the two roads diverged. Now that we have traveled a distance up the road of innocence, perhaps we should try the *biologically responsible* path to see where that leads. This path is not likely to branch in the direction of biological responsibility for a benign or moral gene. If we judge homosexual orientation and its accompanying behavior to be benign or morally desirable, then no one is likely to find this ethically interesting. The other branch that combines responsibility and defect is likely to draw attention.

The direction of *biological responsibility for the gay gene defect* would likely include the following suppositions: homosexual expression is socially disruptive, undesirable, and perhaps even immoral; the predisposition to homosexual activity is found in our biological make-up—that is, in our genes; and we inherit these genes from our mothers who inherited them from their forebears going back to who knows where in the history of the human race. Our genes dispose us to certain forms of behavior. In this case, our genes dispose us to desire and, although desire itself is not sinful, actions to satisfy this desire are sinful if not downright corrupting to our nature. Although the concept of the gene per se has played very little role in such thinking until recently, the idea that we are born with physical bodies predisposing us to desire has long been known. Ancient scriptures along with Stoic and Buddhist philosophers have long recognized the near imprisoning power of desire—sometimes called concupiscence—that arises within human consciousness and leads us toward self-corruption and toward social strife. Typically such desire is associated with envy, greed, gluttony, jealousy, and especially sexual lust. Homosexual desire could easily fit on this list. There is no equivalent here of the notion that my biological make-up makes me innocent. Rather, the biological make-up with which I was born saddles me—saddles each of us—with a moral hurdle to overcome.

How do we leap this hurdle? One classic answer is grace. Grace comes in the form of a gift from the Holy Spirit, specifically, the power of the will to live morally. Augustine said that good behavior derives from our will to do the good, and our will is "aided and uplifted by the imparting of the Spirit of grace."[29] What some might argue in light of the gay gene, it appears, is that we would enjoin each individual with this or a similar biological predisposition to find supra-biological resources, such as the power of the Holy Spirit, to gain the strength of will to do the good.

To put it brutally, the control and constructive channeling of desire—any desire, homosexual attraction included—requires self-discipline. The very concept of self-discipline presupposes the existence of a self that can engage in discipline. This implies that the self is more than merely the product of the genotype and phenotype that constitute a person's body. It implies further that the self is more than merely the synthesis of genes and environment, more than mere nature and nurture.

Or, to put it another way, it is not a gene that is sexual. It is a person who is sexual. It is not a gene that desires. A person desires. It is a personal self who both desires and who makes the decisions that control and channel his or her desires. In this regard, each of us is a responsible person regardless of which genes we have or do not have. When theologians speak of grace, they speak of a relationship we as persons can have with God. Specifically, the indwelling of the Holy Spirit produces fruits. These fruits are many. Paul lists them in Galatians 5:22: love, joy, peace, patience, kindness, generosity, faithfulness, gentleness, and self-control. The final item on this list of fruits of grace is self-control.

What I am speaking to at this point is not the morality or immorality of homosexual expression. Rather, I am speaking to the location of responsibility for genetically influenced behavior. When it comes to sexual desire, secular society and religious communities presume that as personal selves we have the capability and responsibility for channeling our sexual expression in nondestructive ways. One meaning of grace is that we are not left alone in pursuing what is good. The power of the Holy Spirit is available to aid us.

Original Sin Again?

Returning to what we said earlier regarding Augustine and original sin, we can see that the *biological innocence with the morally benign gay gene* road mapped above does not lead us to Augustine. Rather, it winds through the Pelagian neighborhood toward individualism. In concert with both the Pelagians and Augustinians this position tacitly affirms that nature is essentially good, or at least good enough. Yet the innocent-benign position is much more individualistic than Augustine could be. The innocent-benign position presumes that *my* nature is essentially good. Nature begins with the genes of the individual gay man, and the appeal to innocence depends on the state of nature at which each one of us finds ourselves at birth. The appeal here is not to the state of nature as a whole, to the human predicament as a whole. This is where the Augustinian position parts company. Augustine, in contrast, begins with the idea of the whole human reality. In Adam all die, whereas in Christ all are made alive. What you or I inherit as individuals from our parents is an inevitable part of the entire history of the human race with sin. Original sin, among other things, marks our identity in unity with the rest of humanity.

If we try to apply the Augustinian notion of original sin to the present discussion, then it would seem that the gay gene would find its place in the larger description of a human condition that includes all of us. The homosexual, heterosexual, and the homophobic predispositions would together constitute signs of a fallen human nature, a historically specific form of human nature in which desire threatens to cause psychic and social strife. The responsible or Augustinian road would ask the bearer of the gay gene to admit what all human beings are asked to admit, namely, we have been born with an inherited predisposition toward sin and that our ethical mandate is to strive for a life of love that transcends our inborn desires. Whether a given individual has the gay gene, the hetero-sexual gene, or the homophobic gene, the ethical mandate applies to all.

Gays and Lesbians in the Churches

Few issues rip and shred the heart more painfully than the culture war taking place within the church and among theologians regarding the place of gay and lesbian Christians. On the one hand, theologians with integrity must wrestle with biblical texts that appear to condemn homosexual expression as sinful. Munich systematic theologian Wolfhart Pannenberg, for example, reviews terrifying texts such as Leviticus 18:22 where one man lying with another is called an "abomination." Leviticus 20:13 pro-nounces such behavior as worthy of death. In Romans 1:27 Paul regards homosexual relations as turning away from God; and in 1 Corinthians 6:9–11 he judges it along with unchastity, adultery, idolatry, witchcraft, drunkenness, theft, and robbery as excluding one from the Kingdom of God. Pannenberg concludes that "In the entirety of the biblical witness, therefore and without exception, homosexual practice is determined to be a mode of behavior in which the turning away of humanity from God is blatantly expressed."[30] He adds that the church must accept the reality that deviations from the biblical norm are common enough, but that these are exceptions that prove the rule. "The church must approach the per-sons concerned with tolerance and understanding, but she must also call them to repentance. She cannot surrender the distinction between the norm and conduct that departs from it. Here stands the boundary for any Christian Church which knows itself bound by Scripture."[31]

On the other hand, many church leaders weep over the emotional and spiritual debris left in the wake of such biblical interpretations. Pannen-berg is a Lutheran. Other Lutherans wrestle with the same texts but find their stands on church life softened greatly by compassion. On March 22, 1996 the Conference of Bishops of the Evangelical Lutheran Church in America issued "An Open Letter" addressed "To gay and lesbian mem-bers" of this church body. This open letter assumes that homosexual people are Christians in good standing, not excluded from the Kingdom of God. "We all live with the pain of a church that experiences sharp dis-agreements on some issues," the bishops wrote; "Yet we walk beside you and we value your gifts and commitment to the Church." Citing a previ-ous national assembly, the letter said that "gay and lesbian people, as individuals created by God, are welcome to participate fully in the life of the congregations of the Evangelical Lutheran Church in America." Recalling what we said earlier regarding the classic Christian commit-ment to the goodness of God's creation, this letter presumes that homosexual persons belong to the good creation. It presumes biological innocence.

The bishops also called for repentance. Yet, whom did they call to repent? Homosexual Christians? No. Christian congregations and the wider society. "We repudiate all words and acts of hatred toward gay and lesbian persons in our congregations and in our communities, and extend a caring welcome for gay and lesbian persons and their families."

Recognizing that the open hand of welcome does not in itself solve the theological issues surrounding the interpretation of scripture, the bishops invited gay and lesbian church members to join in the long hard exegetical study that the future might require. "We invite gay and lesbian persons to join with other members of this church in mutual prayer and study of the issues that still divide us, so that we may seek the truth together. We are determined, despite our differences, to maintain the unity of the Spirit in the bond of peace (Ephesians 4:3)."[32]

The theology of Wolfhart Pannenberg, which we will look at in more detail later, embraces a vision of the future Kingdom of God in which unity is the grand theme: unity of God with us, of us with one another, and of the human race with nature. It would seem that this eschatological vision of human unity could carry Pannenberg from his biblical interpretation more in the direction followed by the bishops, more toward sharing rather than setting this boundary for church fellowship.

To be conscientious and compassionate and remain in community while the wrestling continues is difficult, yet this seems to be the church's vocation for the near future. The ethical struggle will go on for some time. Will the reported possible discovery of a gay gene help solve the theological and ethical problem here? Sadly, no. The scientific fact does not itself determine the direction of the ethical interpretation of that fact. More than scientific information about nature is needed to establish an ethical norm or solve a dispute over ethical norms. Science provides no shortcuts. We must still employ biblical authority, theological interpretation, common sense, sound judgment, and most assuredly compassion and companionship.

Conclusion

My task in this chapter has been to draw a map depicting possible directions that ethical thinking might follow in response to the new genetics in general and the apparent discovery of the gay gene in particular. My task has not been to sell tickets taking us to only one ethical destination. It is not crystal clear to me just what direction gay rights advocates should go, even though the path of biological innocence seems to lead us to helpful insights. What does seem clear to me is that the point of departure given us by current genetic science does not in itself determine our ethical destination. What the genes tell us leaves us immediately at a point where the road forks, requiring us to choose between the path of self-proclaimed innocence, on the one hand, or acceptance of responsibility for what we have inherited, on the other. Although the innocent road may seem to be the obvious one for us in a modern liberal culture to take, perhaps we should pause long enough to give serious consideration to where the responsibility road might take us. It might take us back to the region of original sin and the notion of inherited responsibility, to a consciousness that we share the human predicament with the whole of the human race throughout the ages, and that this unity in sin foreshadows a greater unity in grace.

For those looking for a scientific justification of their existing position on gay rights or even for a red line on the map marking the best road to the land of ethical correctness, this chapter may be disappointing. A theological answer to an ethical question may be understandably disappointing. Sorry. If the reader insists on ending up in an ethical location, then simply return to the map and let your finger trace the direction you wish to take. As you do so, keep in mind that this is a direction you are choosing. It is not a direction genetic science has compelled you to follow.

< FIVE >

Should We Patent God's Creation?

Knowledge as power drives toward ethical dilemmas, and both
knowledge and power drive toward searing questions of meaning.

—Langdon Gilkey

During the weeks leading up to May 18, 1995, well-intentioned religious
leaders prematurely jumped into water over their heads. While thinking
they had a lifeguard watching over them, they in fact were being dragged
out to rough seas. These rough seas are the controversy over gene patenting. The apparent lifeguard is the Foundation on Economic Trends and its
president, Jeremy Rifkin. The rough seas are the knotty intellectual and
ethical debates regarding scientific discovery, the invention of new life
forms, the development of genetically based medicines, and the commer-
cialization of genetic engineering.[1]

During the months following two scientific organizations, the Human
Genome Organization (HUGO) and the Council for Responsible Genetics
(CRG), took stands against genetic knowledge and life-form patenting.
The American Association for the Advancement of Science (AAAS) is
attempting to broker the conversation among scientists, the United States
Patent and Trademark Office (PTO), the biotech and pharmaceutical
industries, and religious leaders. President Bill Clinton agreed to appoint
a National Bioethics Advisory Commission (NBAC) to take up gene
patenting as one of its two principal issues, the other being the rights of
human subjects in research.[2]

Now, what happened on May 18, 1995? And why might this be important theologically? Why might it be important for public policy in America?

May 18, 1995 was the date of a press conference in Washington, D.C. in which spokespersons—who claimed to have collected signatures of nearly 180 individuals known as "religious leaders" representing eighty different faiths or denominations such as Hinduism, Islam, Buddhism, Judaism and Christianity—called for a ban on the patenting of human genes and genetically engineered animals. Press leaks led to a front page story in which the *New York Times* described the event as "a passionate new battle over religion and science."[3] Along with Jeremy Rifkin, the press conference, named the "Joint Appeal Against Human and Animal Patenting," included Rabbi David Saperstein, Director of the Religious Action Center of Reform Judaism; Abdurahman Alamoudi, Executive Director of the American Muslim Council; Wesley Granberg-Michaelson, Secretary General of the Reformed Church in America; Richard Land, Executive Director of the Christian Life Commission of the Southern Baptist Convention; and Kenneth Carder and Jaydee Hanson of the United Methodist Church. "By turning life into patented inventions," said Rifkin, whose Foundation on Economic Trends orchestrated the event, "the government drains life of its intrinsic nature and sacred value."[4]

The revolt against playing God has hit new heights in rhetorical flourish now that preachers have entered the patenting controversy. "Marketing human life is a form of genetic slavery," Richard Land was quoted in papers across the country. "Instead of whole persons being marched in shackles to the market block, human cell lines and gene sequences are labeled, patented and sold to the highest bidders." Land added a prophet-like judgment against playing God in the laboratory: "We see altering life forms, creating new life forms, as a revolt against the sovereignty of God and an attempt to be God."

This event marks a point of public meeting between the religious and scientific communities, a meeting that quite unfortunately has the appearance of a battle. Yet it provides a doorway into some of the theological and ethical conference rooms within which serious assessment of genetic possibilities can be discussed. To open this door we will analyze the thinking of those sponsoring this event, examine the patenting controversy itself, and tease out some of the theological and ethical assumptions. It will be my judgment that our religious leadership has confused the gene myth with serious science. By reacting to the myth at the cultural level, this religious coalition has failed to think through its own theological assumptions; and it has failed to offer a constructive ethical contribution based on an informed understanding of the science involved. I will offer eight recommendations for public policy produced by what I hope will be a more sober theological and ethical reflection on the gene patenting controversy.

What Did the Statement Say?

Let us begin with the actual statement as signed by the religious leaders. Here it is.

> We, the undersigned religious leaders, oppose the patenting of human and animal life forms. We are disturbed by the U.S. Patent Office's recent decision to patent human body parts and several genetically engineered animals. We believe that humans and animals are creations of God, not humans, and as such should not be patented as human inventions.[5]

The first problem with the statement is that it is vague and inflammatory, serving only the cause of Jeremy Rifkin while borrowing a baptism of prestige from the reputations of honored religious leaders. The statement fails to make clear what the Patent and Trademark Office strives to make clear, namely, it does not grant patents to natural phenomena, such as human beings or animals or body parts in any ordinary sense. It grants patents to human inventions that are novel, nonobvious, and useful.[6]

Rifkin, famed for his outspoken resistance to progress in biological research and medical technology, tacitly advocates a naturalist or vitalist philosophy. As I mentioned in an earlier chapter, he describes his mission as a "resacralization of nature."[7] His position is clearly at odds with those religious traditions, such as Christianity and Judaism, which believe in a transcendent and holy God who created the natural world, loves it, and asks that the human race, created in the divine image, strive to make the world a better place. These traditions hold that the creator God is sacred, not the creation.[8]

The ethics that result from these respectively different metaphysical positions is significant. Jews and Christians hold that we humans should be good stewards of our God-given creativity, and out of love for neighbor pursue, among other things, work for the development of better ways to relieve human suffering and improve the human experience. The Rifkin position implies that nature prior to human creative intervention is sacred and should be left alone. This severs the need for pursuing medical research and development of therapies that could relieve human suffering and improve the health of the human race. This is the issue on which the patent debate should focus: How can patents help or retard the development of genetically based therapy for cancer, heart disease, Cystic Fibrosis, Alzheimer's, Huntington's Disease, Williams Syndrome, and countless others? In effect, the religious leaders have unnecessarily cut themselves off from making a contribution to this central concern.

How the Hoodwink Happened

How did it happen that so many otherwise thoughtful theologians and leaders of different religious traditions get hoodwinked? Probably because they read their faxes. A letter of invitation dated February 3, 1995 was sent out from the General Board of Church and Society of the United Methodist Church over the names of Kenneth Carder and Melvin Talbert. Carder, the Nashville Area Bishop, chaired the United Methodist Genetic Science Task Force which produced an excellent study document in 1992. The letter was couched to startle the reader by alluding to the "indiscriminate use" of genetic engineering techniques as "a very real threat to the dignity and integrity of human life." Any conscientious religious leader would want to protect the dignity and integrity of human life, of course. Sounding the gong of urgency, because allegedly the patenting of human material has already begun, the letter included the following material:

> In 1991 the Patent and Trademark Office (PTO) granted patent rights to a California company for commercial ownership of human bone marrow "stem cells" (stem cells are the progenitors of all types of cells in the blood). The PTO had never before allowed a patent on an unaltered part of the human body. Many in the science community were stunned and outraged by the patent. Peter Quesenberry, medical affairs vice chairman of the Leukemia Society of America commented, "It really is outlandish to believe you can patent a stem cell. Where do you draw the line? Can you patent a hand?"

By using such terms as "unaltered" and the singular "a stem cell," these sentences are intended to convey the impression that the Patent and Trademark Office has granted a patent to some unnamed company for a naturally occurring phenomenon in everyone's human body, namely, the appearance of a stem cell. This sounds like a call to arms, a call to rise up against the PTO for granting exclusive rights to something that is natural, human, and—as the letter states later—belonging to God.

Now, just what 1991 event might Carder and Talbert be referring to here? What is that unnamed California company? Did it happen just the way the two Methodists describe it? Where did Carder and Talbert get their information?

Let us answer the last question first. The above cited paragraph is taken almost verbatim from the pages of a book written by Andrew Kimbrell, *The Human Body Shop*.[9] Kimbrell is policy director of the

Foundation on Economic Trends. Jeremy Rifkin wrote the foreword for the book. Kimbrell says more: "What makes the patent remarkable, and legally suspect, is that the patented cells were not any form of product or cell line. They had not been manipulated, engineered, or altered in any way. The PTO had never before allowed a patent on an unaltered part of the human body."[10]

The unnamed California company in fact has a name, SyStemix. I contacted the former SyStemix CEO, Linda Sontag, along with current executives John Schwartz, Iris Brest, and Wendy Hitchcock. I asked if these materials accurately described the patent and its significance. By no means! It is misleading to use the term "unaltered" or to deny that SyStemix has produced a product or cell line. Here is the SyStemix position:

> The patent in question is Patent Number 5,061,620, Human Hematopoietic Stem Cell. Despite the name, however, the patent does not cover single stem cells. Rather, it covers "a cellular composition" whose characteristics are described in detail—basically, a nearly pure collection of pluripotent hematopoietic stem cells— and a method for obtaining the composition. Both the method and the composition were novel, nonobvious, and met all the standard requirements for patents. The patent is not "on an unaltered part of the human body"—indeed, the composition does not exist in the human body or anywhere else in nature. Stem cells in the body are relatively rare and are widely dispersed. The patent does not give SyStemix any rights to anything in the body of any person. The patent does not have anything to do with the physical integrity of an individual or with the ownership of any bodily organ.

What SyStemix achieved was to invent a machine and a process for harvesting stem cells from the blood stream and to create a population of cells outside anyone's body that does not exist naturally. The potential value of this process is staggering. If clinical trials support the theory, medical scientists will be able to place in the bone marrow of cancer patients a purified population of life-giving stem cells that will be uncontaminated by cancer. The import for leukemia and AIDS therapy is enormous. Relevant to the question at hand: no such nearly pure population exists naturally. It has taken highly sophisticated technology to manipulate what nature has bequeathed in order to make this product. Hence, the SyStemix patent has two parts, one for the process and one for the isolated stem cell population.

Is this precedent-setting? No. It is consistent with earlier precedents regarding the purification of natural substances. For example, vitamin B_{12} is found naturally in minute quantities in the livers of cattle and in certain other microorganisms. What we know as the vitamin pill version was patented, and the patent was upheld in court because the purification process yielded a product not found in nature. The pill gives us vitamin B_{12} in abundant supply, relatively inexpensively, free from toxic contaminants, and amenable to control of potency and dosage. Even though vitamin B_{12} occurs in nature, human intervention was necessary to create a new and useful composition of it. The SyStemix patent on the purified population of stem cells follows this precedent.

Knowingly or unknowingly, Carder and Talbert employed misleading information in order to rally religious troops to fight a battle in the army of the Foundation on Economic Trends.

This is certainly how the biotech industry sees it. Carl Feldbaum, president of the Biotechnology Industry Organization, sees Rifkin using the clergy as a cover for an "unusually cynical attack." Even though Rifkin seems unwilling to change his opinions, he continues, the clergy are simply insufficiently informed and could be educated. The clergy need to learn that "Having a patent doesn't mean owning life; it's an ability to develop some aspect of it commercially." The task of the industry is to educate the clergy so they see the connection between patents and the development of new drugs to combat disease and relieve human suffering. Once informed, says Feldbaum, "I think they'll modify their views."[11]

The unfortunate consequence of this event is that our religious leaders may have cried "Wolf!" on the patenting controversy and thereby may have lost credibility with those in the scientific, medical, legal, and commercial communities. Credibility is important here, because serious ethical issues of considerable consequence will confront our society over the next half decade; and our society could benefit from our religious leaders should they take the lead. If we have squandered our moral capital following a pied piper, then we may have nothing left to spend when the issues become more costly.

Other Pressing Issues

What are those ethical issues that genetics research will create or exacerbate? First, and deserving immediate attention, is the threat of genetic discrimination. Because medical care is connected to private insurance and to employment, the development of tests for genetic predispositions to disease—recently discovered are the genes responsible for Huntington's disease, Cystic Fibrosis, Alzheimers, inherited breast cancer, and an estimated 5,000 more to come—we can expect thousands if not millions of people to discover suddenly that they are uninsurable and hence unemployable. This will create a new underclass—a genetic underclass—to add to the 40 million people already inadequately covered by medical insurance. The 1996 Kassebaum-Kennedy health insurance reform bill prohobits insurance companies from counting genetic information as a pre-existing condition. This will discourage insurance companies from denying access to persons who are genetically predisposed to certain diseases. With luck, state governments will follow suit. If religious leaders want to show ethical concern regarding the implications of genetics, then here is a problem crying out for immediate attention.

Genetic discrimination will be closely followed by numerous other problems such as selective abortion—that is, as prenatal testing expands we can forecast wholesale aborting of fetuses because they do not meet genetic standards. This will be more than a mere moral issue with the parents; we can expect the insurance industry to demand abortion for potentially expensive babies. The result is likely to be market-driven eugenics.

These problems will be followed by the knotty question of moral and legal responsibility for antisocial behavior for which some individuals have a genetic predisposition. Here we have been asking, if it can be shown that some persons are genetically predisposed toward alcoholism, aggression, or violence, will society consider them guilty or innocent? What will the courts do when those having committed crimes claim innocence on the grounds that "my genes made me do it"? And how will the logic of responsibility for our behavior help us to understand the implications of the 1993 discovery of the gay gene?

These ethical conundrums will be followed by still more, such as the controversy over somatic therapy versus germline intervention. Should we engage in genetic engineering that will influence the evolution of future generations?

In short, if religious leaders are genuinely interested in taking up ethical issues facing our society over genetic research, then the agenda will be long, serious, and complicated. A quick endorsement of Jeremy Rifkin's hysteria over patenting is no short cut.

Finally, the scientific community along with the legal community is already wrestling with the patent issue. Scientists might even appreciate some considered deliberation and ethical insight offered by religious leaders. This is not the moment to blast scientists with a hostile barrage of partially informed sanctimony.

The Theology of the Joint Appeal against Human and Animal Patenting

The statement signed by the 180 or so religious leaders included this creedlike confession: "We believe that humans and animals are creations of God, not humans, and as such should not be patented as human inventions."[12] Let us explore this, asking just what might be meant by the theological and ethical logic employed here. Could it mean that because humans and animals are creations of God and not something we humans have created, it follows that they should not be patented? If so, the category of *creation* here sharply distinguishes between something God creates from something we humans create. To what might this refer? The inherited DNA sequences including genes that we find in our cells already exist in nature. We did not create DNA through scientific or technological invention, to be sure. Is it this quality of being inherited rather than invented that qualifies it as God's creation?

What about genetically engineered animals? Should we presume that the statement is referring here to the Harvard mouse and other transgenic animals being bred for medical purposes? These are life forms that hitherto never existed in evolutionary history. They represent new genetic combinations engineered for specific purposes, usually for conducting medical experiments. The DNA sequence of these animals did not exist in nature. They are the result of human intervention. Now, let us ask: Are these too creations of God? Perhaps not according to the statement, because they are human creations. Or, could the statement read: genetically engineered animals really are creations of God and the human scientists who patented them are deluded in thinking that they, not God, did the actual creating? This would be silly, of course. Yet, such silliness is unavoidable with such a statement. A charitable interpretation would dub the statement unclear. A literal rendering would interpret the statement to be sharply separating two forms of creation: divine from human.

Theologically, just how sound is it to separate so sharply between what God creates and what human invention creates? If we make such a distinction, then how do we identify what God creates? Should we presume that something found in nature that has not yet been touched by human technology constitutes a divine creation? Does it mean that as soon as human creativity alters its natural state that it leaves the realm of divine creation and enters the realm of human creation? Does it mean that God creates only part of reality and that human technology creates a different part?

Numerous theologians in our era would avoid separating divine from human creativity at this point. If we think of God's creative work in terms of both creation out of nothing (*creatio ex nihilo*) and continuing creation (*creatio continua*), then two relevant notions follow. First, the sheer existence of the natural world to which we in the human race belong is totally dependent on the creative act of God. Without God's creative work, nothing would exist. In this sense, all of nature including human nature belongs to the realm of God's creation. Nothing lies outside God's creation.

Second, creation is a process that is still going on. One might even include evolutionary history in the story of God's ongoing creative work. One might further include the evolutionary development of the human genome with its resulting intelligence and creativity—the ability of the human being to alter itself—as an additional chapter in the story of nature. At this point divine creativity and human creativity enjoy some overlap. They are not sharply separated. This two-dimensional understanding of creation is codified in the anthropological proposal of Philip Hefner, where he refers to the human being as God's *created cocreator*.[13] This line of reasoning would suggest that it is misleading to argue that technological intervention into the cell line of a life form, even a human life form, lies outside the realm of God's creation.

Turning to the ethical principle derived from this theology of creation, the statement concludes that humans and animals ought not be patented. What is being referred to here? Let us ask about humans first. Does it refer to individual human beings? Certainly no humans could legally be patented. The principle of human dignity would proscribe this. A patent permits its owner to make, use, or sell its product, and such treatment of a human being would violate principles of liberty guaranteed by the Thirteenth Amendment to the U.S. Constitution. To my knowledge, no one has ever seriously tried to patent an individual human being. This is not a category for patent consideration.

What about groups of human beings? Perhaps a race or a clan or a family? Again, noting the proscription against slavery in the U.S. Constitution, this is simply not a category that fits the patenting discussion.

Adding confusion here is the phrase, "body parts." This phrase comes to us from Kimbrell's inflammatory book, *The Human Body Shop*. Chapters from Kimbrell's book were photocopied and mailed out to entice religious leaders to sign the statement. Kimbrell makes a speculative but provocative remark. He says that under the Patent and Trademark Office's 1987 ruling, "a genetically engineered human kidney, cornea, arm, or leg, or any other body part might well be patentable." Kimbrell knows that this does not refer to kidneys or arms on existing people. He also knows that no such patent applications are being filed. He grants that this is a wild extrapolation based on what is in fact happening, namely, "the patenting of human cells, genes, and other biological materials."[14]

So, perhaps we should ask: Could the statement by the religious leaders be referring to the patenting of human DNA sequences? Human DNA sequences are not human beings, to be sure; but, given the context and the possible influence of Andrew Kimbrell, perhaps this is what the signers had in mind. It would be odd to think of DNA as a body part; but we should explore the possible logic here. Might the statement be referring to the entire human genome—that is, the total number and function of all human genes? No one knows the entire genome yet. This is what the Human Genome Project, expected to last into the next century, will try to learn. Because this is a worldwide cooperative scientific enterprise, it is inconceivable that one scientist or one company or one government could apply for exclusive patent rights to make, use, or sell the entire human genome.

What about portions of the nucleotide sequence in the human DNA, perhaps part of a gene or even an entire gene? Now we are getting closer to reality. Yes, knowledge of some DNA sequences and some genes have been granted intellectual property rights by the patent office. We should note that genes as they exist in human nature are not patentable. What is subject to patent consideration are genes that have been altered—that is, new forms of genes produced through human intervention. Later we will discuss the distinction between *discovery* of existing genes and the *invention* of new genetic forms when examining the patent controversy of the 1990s.

Before turning to the patent controversy itself, we should note that the industry reaction to the Joint Appeal is fascinating. It is fascinating because the Joint Appeal may have inadvertently accomplished something significant, namely, it has drawn nonreligious interests into a public debate over theological matters. "Who Speaks for God?" is the title of the publisher's commentary in an industry newsletter, *BioCentury*. Its subscribers include the professional and investment community such as bankers, brokers, pension managers, and CEOs who are concerned about bringing new medical products to market. "However profane it may seem," says *BioCentury*, "intellectual property protection [through patents] is what makes it possible for the inventor and developer of a technology to be rewarded for risking the hundreds of millions of dollars and decades of time required to bring new drugs to market."

The loss of an exclusive market for a new drug, which a patent provides, would be a strong disincentive to rally the financial support to pursue the research. Apart from the enormous amounts of capital raised in the private sector, no other agency including universities or the federal government could make up for the lost support. The result of denying patents would be "a diminution of novel treatments for poorly treated or untreated diseases." *BioCentury* then goes on to introduce a threat to the health of a very large medical market, namely, the baby boomers.

> Given the time required to develop truly new treatments, even a delay of a few years caused by this patent uncertainty would mean that treatments for diseases of aging may reach the market too late to benefit the oldest of the baby boomers, and certainly too late to benefit the parents of boomers with Alzheimer's and other heart rending conditions. Given the choice between prayer and new therapies, we're certain many of the boomers would take the therapies and make their separate peace with God.[15]

Perhaps the temptation to introduce a note of cynicism and threat was irresistible: either give us patents or we'll give you Alzheimer's! Without patents, all you have left is prayer!

In the statement that Kenneth Carder actually read on May 18, 1995—to be distinguished from the letter he coauthored that was reprinted in numerous places, including *BioCentury*—the bishop made a salutary and forceful point. "The patenting of genes, the building blocks of life, tends to reduce it to its economic worth. Life becomes a commodity whose value is determined by its commercial value." Despite the confusion over the relation between genes and life, Carder's warning to guard against reducing life to commercial value is well taken. It should be heeded. The *BioCentury* reaction is noteworthy. Seeking rapprochement between the industry and the clergy, the newsletter offers these olive branches:

> Perhaps we could agree that a cell line or a single gene aren't life forms in the same way that an entire human being is, and that ownership of these doesn't constitute a threat to the sanctity of life.
>
> Maybe we could agree that ownership of a genetically engineered animal line is not a license to disrespect life any more than ownership of an individual animal is.[16]

The relation between the dignity of living beings and the patenting of intellectual knowledge regarding gene sequences and cell lines is a delicate one, and the Joint Appeal may have done the public a service by alerting us once again to the danger of commercializing life.

In summary thus far, human beings per se, or parts of human beings as found in nature, are not now nor ever have been subject to patenting claims. Yes, the Patent and Trademark Office has granted intellectual property patents on DNA sequences including genes, but only altered genes.[17] The question of patenting knowledge of DNA sequences is well worth a public debate; but this statement signed by religious leaders lacks sufficient precision to enter into such a debate. To the background and foreground of that debate we now turn.

The Controversy over Patenting DNA Sequences

The patent controversy of the 1990s was sparked by the initial filing in June 1991 for patent property rights on 337 gene fragments, and a second filing in February 1992 on 2,375 more partial gene sequences by J. Craig Venter. At that time Venter was pursuing research at the National Institute of Neurological Disorders and Strokes of the National Institutes of Health (NIH). His aim was to locate and sequence the 30,000 or so complementary DNAs, or cDNAs, from the human brain.

Here is Venter's method, built on a significant though deceptively simple insight. The task of the Human Genome Project is to sequence the entire three billion nucleotides in the DNA and to locate where on the DNA the genes are sited. Relatively speaking, only a small portion of the DNA functions as genes, about 3 percent. The nongenetic material has been affectionately labeled "junk DNA." If one wants to find only the genes, thought Venter, then why bother with plodding through all the junk DNA? Noting that only the genes, not the junk DNA, code for proteins by creating messenger RNA (mRNA), Venter set his focus on mRNA. He began making sturdier clone copies of the otherwise fragile mRNA; and these stronger and analyzable copies he called cDNAs. By sequencing only the cDNAs he could be assured that he was gaining knowledge of actual genes; and by focusing the research this way, he brought the price of sequencing down dramatically.

What we have then are clones made from messenger RNAs, meaning that they represent a part of the coding region of genes on the DNA chain. By sequencing a short stretch of cDNA clones—about 300 to 500 bases and not necessarily the entire gene—Venter created what he called an "expressed sequence tag," or EST. Venter had begun using automatic sequencing machines to the limit of their capacity and was churning out 50 to 150 such tags per day. These cDNA exist outside any human's body, to be sure; yet they constitute a copy of existing genomic material.

A preliminary U.S. Patent and Trademark Office ruling in the fall of 1992 denied the applications on the grounds that gene fragments could not be patented without knowing the function of the gene. Then NIH director Bernadine Healy pressed for an appeal. When Harold Varmus took over the helm from Healy he withdrew the applications, saying that patents on partial gene sequences are "not in the best interests of the public or science."[18] This volleyed the ball into the private sector where similar patent applications have been filed. Some have been filed by Venter, and his private sector colleagues, after he left NIH. Having garnered $70 million in venture capital to start a private biotech company, Venter now heads The Institute for Genomic Research (TIGR).[19] The result has been controversy with numerous hotly debated questions.

What are the hotly debated questions? Initially there were two: Should patents be granted for knowledge of gene sequences at all? And, if so, should they be granted to a government-funded agency such as NIH or only to the private sector? The first question regarding the patentability of genetic knowledge can be broken down into three subquestions. We will turn to these now and then take up later the issue of government patents.

First, does successful cDNA sequencing count as patentable knowledge about the genes themselves? Venter admitted that even though he could tag a cDNA sequence he had no idea what its function was, unless it belonged to a sequence from a gene whose function he already knew. James D. Watson, then head of the National Center for Human Genome Research (NCHGR) at NIH vitriolically opposed this rush to patenting, decrying the overvaluing of what was being accomplished.[20] Simply sequencing a short piece of an unidentified clone with an automated sequencing machine is a "dumb, repetitive task."[21] What remains to be done, and what is decisive, Watson said, is to interpret the data so that we learn exactly what function each gene performs. Similarly, the Gene Patent Working Group, an interagency committee set up by the White House Office of Science and Technology Policy (OSTP), declared that ESTs are merely research tools and should not be granted patent protection that belongs to knowledge of the complete gene sequence with its function.[22] In sum, critics argued that patenting at this early stage was premature.

Second, should this particular knowledge be granted a patent in the form of intellectual property protection? To patent, an invention must meet three criteria: it must be novel, nonobvious, and have utility. Certainly Venter's application was novel, because the newly identified genes seem not to have been sequenced before. However, it is less obvious that it is nonobvious. Venter made use of existing technology; he did not invent it. Nevertheless, Venter contributed an insight that led to unprecedented speed in scientific searching ability. This must be worth something, his sponsoring lawyers argued. Even so, this does not speak to the toughest hurdle: utility. Venter's genes were dubbed "naked," meaning that their function was unknown. Until the function could be known, no movement could be made toward developing medical or other benefits.[23]

The third subquestion is more broadly philosophical: Should intellectual knowledge regarding natural processes in principle be patentable? Does witnessing an existing natural phenomenon in itself warrant patent protection for the witness? Recalling that Sir Edmund Hillary and Tenzing Norgay were the first mountain climbers to see the top of Mount Everest on March 29, 1953, one might ask by analogy: Should Hillary and Norgay have been able to patent Mount Everest? No, would be the answer someone like Justice Douglas would give. Writing for the majority in the 1948 U.S. Supreme Court case of *Funk Brothers Seed Co. v. Kalo Inoculant Co.*, he wrote: "patents cannot issue for the discovery of the phenomena of nature. . . . [Such] are manifestations of laws of nature, free to all men and reserved exclusively to none."[24]

When trying to apply the Douglas principle to the intellectual property issue regarding patenting partial gene sequences, we should distinguish between discovery and invention. This is what patent attorney Kate Murashige does. The human genome is not invented, she holds.[25] Rather, the make-up of the human genome is something scientists are right now in the process of discovering. Like Mount Everest, it is something already there. The discovery might be new, but the phenomenon of nature in itself is not novel.

Are cDNAs a natural phenomenon or a human invention? The cDNA is not a gene per se. Rather, it is a copy version of a gene with the introns edited out. It does not occur naturally. It is coded into messenger RNA by the process that reads the raw cellular DNA. This fact leads to an interesting double-mindedness on the part of science historian Daniel Kevles and geneticist Leroy Hood. On the one hand, they argue: "since it can be physically realized by a devising of human beings, using the enzyme reverse transcriptase, it is patentable." On the other hand, Kevles and Hood are troubled. "If anything is literally a common birthright of human beings, it is the human genome. It would thus seem that if anything should be avoided in the genomic political economy, it is a war of patents and commerce over the operational elements of that birthright."[26]

In my judgment, the question regarding the novelty of cDNAs is not settled. This is important because, even though NIH withdrew the original Venter applications, nearly 1,200 so-called "gene patents" have been issued as of the date of this writing, most in Japan and the United States with relatively few in Europe. Tens of thousands more applications are pending in the United States. It seems to me that at least three ambiguities lie at the very axis of the surrounding controversy. As long as these ambiguities go unresolved, the controversy will go on. The question is this: To what does "gene patents" refer? The answer: it refers to different things.

The first ambiguity has to do with two types of patents, *process patents* issued for the process by which a gene is isolated versus *composition of matter patents* issued for the physical product of the process. Currently gene patents are issued on both. The United Methodist Church approves of process patents but opposes patents on genes themselves understood as a composition of matter. The industry and the PTO prefer the latter because it is easier to defend composition patents in court.

The second ambiguity has to do with whether the gene patented is discovered as already existing in nature or the product of human invention. Because the actual gene to be patented is a cDNA which is the result of a laborious process of copying the original DNA, some patents have been issued on cDNAs (not on the original DNA of a particular human being). Yet, the term "gene patent" fails in itself to distinguish between the two, giving the false impression that a patent on the latter is a patent on the former.

In fact, the false impression just may be the correct impression, as the next ambiguity attests. The third ambiguity has to do with whether the gene should be understood as a unit of information or as a chemical structure. Many molecular biologists think of the gene as a unit of information, as a code. This code directs the DNA to produce proteins. When thought of as an information code, the DNA sequence is identical whether found in the original DNA or in the cDNA. A patent on the latter applies to the former. Intellectual property rights to cDNA apply to the original genome. On this basis, one could rightly complain: a patent on cDNA affords rights regarding what previously existed in nature. It violates the Douglas principle. The PTO, in contrast, seems to be assuming that a gene is a biochemical structure. This assumption permits the distinction between genomic DNA and cDNA.

At present the effect of these ambiguities is to encourage applications for gene patents. Scientists engaged in basic genomic research, however, tend to oppose the issuing of patents on raw DNA data gained through producing cDNAs. Their primary argument rests less on the point that such patents violate the Douglas principle. Rather, the primary concern of the researchers is the flow of scientific information.

In 1995 the Human Genome Organization (HUGO) issued a statement opposing the patenting cDNAs because it would impede the free flow of scientific information. "HUGO is worried that the patenting of partial and uncharacterized cDNA sequences will reward those who make routine discoveries but penalize those who determine biological function or application. Such an outcome would impede the development of diagnostics and therapeutics, which is clearly not in the public interest. HUGO is also dedicated to the early release of genome information, thus accelerating widespread investigation of functional aspects of genes."[27]

Andrew Kimbrell offers a more fearful interpretation of what might happen as a result of the Craig Venter affair.

> The entire human genome, the tens of thousands of genes that are our most intimate common heritage, would be owned by a handful of companies and governments. If Venter's applications or those of others are accepted, in a short time a few government bureaucracies and powerful corporation will have a monopoly on the use and sale of all human genes.[28]

In sum, cDNAs may prove patentable on the grounds that they are the product of a humanly devised process of gaining intellectual knowledge. Some patents are now being issued with this justification. A strict rendering of the Douglas principle that things existing in nature are not subject to intellectual property protection would, in my judgment, render such patents unfortunate.

Should the Government Hold Patents?

Now, let us turn to the second large question raised by the Venter affair: Should the U.S. National Institutes of Health itself seek patents and hold the power to license?[29] Past NIH director Bernadine Healy said yes, arguing that NIH patents would best support new product development in the private sector. The NIH motive was not primarily to seek financial return for the government, but rather to gain patent rights and then disseminate the knowledge to promote the development of new products elsewhere. The avowed purpose was to enable the NIH to offer licenses to biotech companies, a process of technology transfer from government to private industry in order to stimulate research into genetically based diseases and the manufacture of drugs and other therapies based on knowledge of these sequences.

Publicly opposing Healy, then NCHGR director James Watson said no, blasting the initial filing by Craig Venter as a land grab, a preemptive strike that would likely promote a worldwide stampede to garner patents on essentially meaningless pieces of DNA. The basic problem, as Watson saw it, was that the NIH patenting policy would foster secrecy among scientists; it would destroy the fragile practice of open sharing of information among scientists around the world.

Watson was right, in my judgment and in the judgment of many scientists. Anger toward the United States flared up around the world. For example, David Owen of the United Kingdom's Medical Research Council (MRC) retaliated by filing hundreds of patent applications, saying that his government wanted an international agreement that no country would seek patents on gene sequences of unknown utility. Until such an agreement could be reached, the British government would file patent applications to protect its own research.[30] French geneticist Daniel Cohen spoke for many critics when he told the press that there were two big problems with NIH patenting: "The first is moral. You can't patent something that belongs to everyone. It's like trying to patent the stars. The second is economic. By patenting something without knowing the use of it, you inhibit industry. This could be catastrophic."[31]

A public fight between Healy and Watson erupted. Watson stomped out of NIH. Healy accepted his resignation. When the Democrats took over the White House, Harold Varmus took over NIH and Francis Collins became director of NCHGR. The NIH patent applications were dropped. Yet, the controversy continues.[32]

The central issue is the fear on the part of scientists that government patents, and in some cases even private patents, will obstruct the free flow of information regarding the state of nature that belongs inherently to the public domain. In part, such a fear is well founded, because some researchers will be tempted to keep what they learn secret until the patent application is well under way.

In another sense, however, the fear is less well founded. Patent law itself requires a rigorous disclosure. The patent makes the knowledge public. A patent gives exclusive right to make, use, or sell a product; it does not keep the knowledge of that product secret. It publishes that knowledge. What the scientific community needs most is not a ban on patents. Rather, what it needs is an experimental use policy—that is, a clear and enforceable policy that exempts scientists from patent restrictions so that they can use patented materials in their research. Such an experimental use exemption currently exists, but many believe it is too narrow. Perhaps a broader exemption from patent restrictions given to researchers will suffice in meeting what scientists see as the need to maintain the free flow of information.

Before turning to the logic of the NIH claim that its own patents would facilitate technology transfer to the private sector, let us examine a line of argument that says the following: patenting, whether by government or the private sector, is unnecessary because public funds are already being used to support the Human Genome Project. Because funds are available for research, it might be argued that patent incentives are not needed to stimulate investment. One could reinforce this by objecting to a double dip: citizens pay first as taxpayers and then as consumers, when they pay royalties to the patent holder for use of a tax-supported invention.

This line of argument is not convincing, at least to University of Michigan Law professor Rebecca Eisenberg. She presents several reasons for supporting private-sector patents even if federal funding underwrites some of the basic sequencing research. First, in terms of actual dollars spent, the amount of money granted by the U.S. government is trivial compared to the amount of private funding for biotechnology research and development. The NIH and DOE budget for the Human Genome Project is roughly $200 million per year. Compare this to an estimated $1.5 to $2 billion invested by private industry. It is doubtful that the U.S. Congress would be willing to compensate the loss in dollars should patent incentives be removed.

Second, federal funding at the moment is limited to mapping and sequencing the human genome and, to some extent, locating disease related genes. The lion's share of focus on producing products such as pharmaceuticals and therapies is taken by the 1300 or so biotechnology companies. Eisenberg concedes that private firms may not need patent protection on DNA sequences in order to draw investment for new product development. No profit can be gained from information about the genome per se. The investment is aimed at processing this information and producing products. This is not enough to decide the issues completely, however. Most of the processes leading to new products are considered standard, or state of the art, and thus unpatentable. We cannot get by by simply distinguishing between the natural DNA sequences and the innovative processes that depend on knowing the DNA sequences. Therefore, she argues, "some form of product patent protection may be necessary, whether the patented products are DNA sequences, recombinant materials incorporating such sequences, or the proteins they produce."[33]

Third, even though the federal government in the past viewed private ownership of inventions made through public funding as contrary to the public interest, this is no longer the case. The express purpose of the patent system today is to promote the utilization of inventions arising from federally supported research and development.[34] The lines between government and private industry are crossed daily, as scientists play "musical labs" and dance from NIH grants to biotechnology firms and back.[35]

This is not a perfect situation. On balance, Eisenberg finds she can support patenting. She writes,

Given the importance of funding from private industry to biomedical research today, it is not clear that patent protection for
DNA sequences is a bad thing. The small amounts of federal
funding projected for the Human Genome Project cannot be
expected to displace private funding in this area. Moreover, even
if public funding were sufficient to generate sequence information
itself, the lack of intellectual property rights in DNA sequences
might undermine incentives for the private sector to support subsequent research to put this information to practical use.[36]

Now let us turn to the NIH policy claim that by obtaining patents on
government funded research it will promote product development in the
private sector. Midwife to the birth of the Human Genome Project was
the belief on the part of the U.S. Congress that a federal infusion of money
into genetic research would enliven the biotechnology industry in the
United States and increase this country's competitive edge in trade over
Japan. The NIH strategy was to obtain intellectual property rights regarding knowledge of the human genome and then license its exclusive
use by private firms who might find research and development too risky
or too costly if this information were left open in the public domain. This
policy of technology transfer via patent licensing was hoped to provide a
way to stimulate new biotech products for worldwide consumption and
to create jobs for U.S. workers while protecting U.S. firms from foreign
competition.

Encouragement for patenting in general is strong. In a government-
funded university research project, for example, if the particular government agency has no interest in filing for a patent, and if the university
receiving the grant has no interest, then the professors involved are
encouraged to step in and claim patent rights. On the assumption that
intellectual knowledge left to the public domain will not be used, this
patenting policy is being pursued in the name of advancing productivity
in American industry.

It is of more than minor significance that even though Washington sees
it this way, American industry does not. Specifically, with regard to the
proposal to patent knowledge of DNA sequences, industry trade organizations such as the Pharmaceutical Manufactures Association (PMA) and
the Industrial Biotechnology Association (IBA) object to NIH patenting.
They prefer to see such knowledge remain in the public domain. Although
a third group, the Association of Biotechnology Companies (ABC) favors
government patenting so that NIH can make money to support its own

research, ABC also urges that such patents be licensed on a nonexclusive basis. These three trade groups fear that NIH patents threaten to block development in the private sector, not facilitate it.

Why are biotech trade groups opposed to government patents on DNA sequencing? There are three reasons. First, government patent rights seem unnecessary, and may in fact turn out to be an annoyance. Knowledge of DNA sequences provides a research tool, not a marketable consumer item. Each company will seek its own patent rights on more complex products anyway; so the obtaining of government dispensed licenses at the research stage is an unnecessary extra burden. Second, the trade groups fear that NIH power to determine who receives and who is denied licenses constitutes excessive government control. The NIH might use this power to control pricing or impose other restrictions unacceptable to research and development companies. Third, there is no guarantee that NIH patents will be effective in protecting a favorable market position for innovative American firms amidst worldwide competition. If in the licensing the NIH insists that the licensed firm employ only American workers or impose other expensive hurdles, then this will drive the product price up and make the firm less competitive.

As to whether or not the NIH should qualify for patent rights on cDNA sequences, Rebecca Eisenberg gives the two reasons mentioned above for denying the applications: (1) the intuitively appealing argument that patents should not be issued for the discovery of things that exist in nature; and (2) they lack known utility.[37] These arguments would apply in principle to any patent applicant, not just the government. She goes on to observe that in the case of cDNA sequences the NIH patenting policy would most likely fail to meet its avowed objective of actually facilitating technology transfer. Even if federal patenting might achieve this end in some circumstances, the U.S. government would be better off here trying to enrich the public domain with this newly acquired knowledge.

What is current research policy? National Center for Human Genome Research director Francis Collins, even though a co-owner of the patent on the gene for cystic fibrosis, is sympathetic to the position previously enunciated by James Watson and HUGO, namely, that for the sake of the free flow of scientific information gene patents should be discouraged. On April 9, 1996, NCHGR at the NIH issued a directive to this effect. It stated that raw knowledge of a genomic DNA sequence, in the absence of additional demonstrated biological information, lacks demonstrated utility. Because it lacks utility, it is inappropriate material for patent filing. Therefore, the NCHGR discourages all its funded researchers from making such applications. This is NCHGR policy, not law. The NCHGR cannot deny the right to file. The only power to enforce this policy is found in the NCHGR's power to consider renewed funding for such researchers.

Now a certain humorous irony enters the picture. Because of the success of the Human Genome Project in mapping the genes—10,000 genetic markers now exist—the time is ripe for accelerated work on sequencing the DNA so we can finish ahead of the scheduled completion date of 2005. The NIH has entertained options for funding someone with an up to date laboratory. In addition to employing the most advanced sequencing technology, this laboratory must also agree to the NCHGR policy against gene patenting. The agreements have been made. The grant decision is now in. The funding will go to The Institute for Genomic Research (TIGR) and its principal researcher, Craig Venter.

Should We Patent Living Beings?

Although it seems obvious, perhaps we should outrightly state: a cell line with a known DNA sequence is not a living being. Even if DNA deserves an appellation such as the "blueprint" of life, in itself it is not life. Nor is it a human being. Nor is it a person. Therefore, the debate over patenting DNA sequences—regardless of which side one takes—is not a debate over patenting life.

Yet, patenting whole life forms is also an issue raised by the Joint Appeal. The religious leaders involved here are not the first to raise their voices against the idea of patenting living beings. The Patent and Trademark Office itself resisted the idea for a long time. This policy seemed to change in 1980 with the Supreme Court decision regarding oil-eating bacteria, *Diamond v. Chakrabarty*.

The policy change began with the 1971 patent application by a General Electric microbiologist, Ananda Mohan Chakrabarty. The task as GE had conceived it was to develop an oil-eating microbe that could be used to clean up petroleum spills on the ocean surface. Chakrabarty fused together genetic material from four types of pseudomonas bacteria to create a crossbreed with an appetite for oil. The rings of bacterial DNA, or plasmids, contained the genetic information to enable the new organism to break down the components of crude oil. The oil-eaters performed admirably in the laboratory, but their fragility made the open seas program untenable. Nevertheless, GE pressed forward even after their initial application was rejected by the PTO.

Why the rejection for a patent? The initial patent examiner offered two reasons. First, because the microbes were "products of nature," and hence not novel. Second, as living organisms they would be unpatentable. The PTO received the examiner's recommendation but altered the evaluation; it concluded that the new bacteria containing the multiple plasmids were not "products of nature." They would pass the novelty test. Nevertheless, the PTO followed through, rejecting the application on the grounds that animate life forms are not patentable.

GE appealed. Processing the appeal took some years, and eventually GE won. The Court of Customs and Patent Appeals (CCPA) ruled that Chakrabarty could patent the oil-eating microbe, saying that the fact that these microorganisms are alive is without legal significance. The PTO appealed, and in 1980 the Supreme Court reaffirmed the CCPA, holding that a genetically altered organism may qualify for patent protection as a new "manufacture" or "composition of matter."[38] Because Chakrabarty's discovery was not nature's handiwork, but his own, it is patentable subject matter even if it is animate subject matter. Underscored in the *Diamond v. Chakrabarty* case is the criterion of novelty; diminished is the criterion of inanimate versus animate.

The story continues. In 1987, after approving a patent for polyploid oysters, the PTO issued a notice alerting the public that it "now considers nonnaturally occurring nonhuman multicellular living organisms, including animals, to be patentable subject matter."[39] One could not patent a human being, however, because "the grant of a limited, but exclusive property right in a human being is prohibited by the Constitution." The reference here is to the Thirteenth Amendment prohibiting slavery.[40]

On April 12, 1988, the PTO approved Patent No. 4,736,866, the "mouse that roared." It was the first of two transgenic mice to be patented by Philip Leder, a Harvard professor. What has come to be known as the "Harvard Mouse" is really a cell line of oncogenes that make the mouse particularly susceptible to cancer, an excellent subject for cancer research. This cell line could in principle be placed in any animal, but mice became the research animal of choice at Harvard.

The loud objections to this 1988 patent make it the "mouse that roared." Animal rights advocates object on the grounds that transgenic animal research involves cruelty to animals. Furthermore, research on animals reveals an arrogant human disregard for the integrity of nonhuman species. We will not discuss an evaluation of these concerns regarding the suffering and dignity of animals, though they certainly deserve a fair hearing. I would like to point out that these objections are aimed less at patent protection and more at the underlying philosophy guiding the research and technology as a whole. Instead, we will examine what all of this bears on our discussion of playing God.

Among the loud objections are those raised by religious voices that say living beings belong to God's creation in a special way, a way distinguishable from inanimate matter. These are the "playing God!" objections—that is, it is an act of human *hubris* to create new life forms for the purpose of medical research that will benefit only the human race, or worse, for the purpose of making a profit in the commercial marketplace.

It is worthy of note that the Joint Appeal Against Human and Animal Patenting does not limit itself to only living creatures as bequeathed to us by the evolutionary process. It includes concern for transgenic creations—that is, new breeds of animals resulting from the intervention of genetic technology. Despite meeting the criterion of novelty, these new animals should not be patentable, say the critics. To patent a living thing is to treat it as a nonliving thing: to commodify it; to commercialize it; and to deny its dignity and integrity. What is being expressed here is a sensitivity to living beings, an affection or even reverence for the life of animals. Regardless of how the patent controversy is finally resolved, animal rights advocates and religious spokespersons do society a great service by keeping us ever mindful of the inherent integrity of living beings and the moral maxim to minimize suffering in this world.

The response of the biotech industry is to minimize the degree of change that actually takes place through engineering new life forms. After agreeing that ownership of a genetically engineered animal is not a license to disrespect life, *BioCentury* goes on to argue that practices such as animal breeding and plant hybridizing have had greater impact on altering life forms than current genetic engineering.

> Humans have bred and developed and owned plant and animal lines since the first dog wandered into camp and the first agriculturist saved last year's seeds for the next spring. (We also would point out that if one is going to object to ownership of animals and human cells and genes, there's no logic to stopping there. Why not continue to prevent ownership of plants, bacteria and viruses, which are no less God's creatures than mammals! Why not ban the eating of meat or the wearing of leather; Death is surely at least as exploitative and immoral as mere ownership).
>
> This is also a good opportunity to explain that much of the genetic manipulation practiced by biotech has long been practiced by Nature (or God), and that much of the genetic manipulation practiced by

biotech is less extensive than the genetic manipulation that occurs in conventional breeding.[41]

These concerns regarding life forms and their alteration, important as they are, are not essentially patenting issues. They belong to a wider discussion about the philosophy of nature and guidelines for pursuing scientific research involving animals and human subjects. With or without patents on transgenic animals, these concerns still deserve public deliberation. Rather than patent laws, perhaps we need guidelines for the conduct of laboratory research, perhaps even outright prohibitions on certain forms of research when it involves inhumane treatment of animals.

What about Cross-Species Genetic Engineering?

Finally, what about mixing human genes and animal genes? What about creating new genetic codes that cross lines between species? Does it dehumanize the human? Does it deny integrity to the animal species? Though these questions sound alarming, in principle they are dealing with nonissues. Genes are genes whether we find them in humans, animals, or plants. Genetically speaking, no sharp lines divide the species. We human beings share 60 percent of our genome with the rats scurrying under our house and all but a percent or two with chimpanzees and other apes in the jungle. Only the tiniest fraction of our genes distinguishes us as humans from the rest of the living world. An even tinier fraction of only a single percent distinguishes us as individuals from each other. Therefore, the creation of transgenic animals crosses no border or violates a "no-trespassing" sign put there by nature to keep us out. Whether we like it or not, we human beings are a part of nature, and nature is us.

Nevertheless, placing distinctively human genes in a transgenic animal raises a rather curious theoretical problem regarding patenting. As we mentioned above, the 1987 PTO ruling rightly excludes the patenting of human beings by citing the prohibition against slavery. Yet, transgenic animals carrying some human genes can be patented, and this gives the patent holder the right to make, use, and sell the animal breed bearing those genes. Is there at least a minimal sense in which something human has been subjected to property rights? Have we taken the first step on to a slippery slope that might find us sliding toward a Brave New World in which genetically engineered humans would be owned and operated by patent holders? Such issues deserve further exploration.

Returning briefly to the May 18, 1995 press conference, the oral statement of Methodist bishop Carder was much more discerning than the signed written statement or even his own letter of invitation. Although

Carder still failed to show the connection between DNA and life—he seemed to presume they are identical—he brought focus to the discussion. "The issue, therefore, related to the patenting of genes . . . is the commodification of life and the reduction of life to its commercial value and marketability." Because life is a gift from God the creator, and because as a gift life has intrinsic value, he argued, the problem with patenting is that it reduces life to its economic worth. Whether life as we find it or as altered through genetic engineering, Carder is concerned that life might become just one more commodity whose value is determined by its commercial value. The Carder caution is well worth attending to.

Concluding Recommendations

Ordinarily one might not think that a connection exists between the interests of religious leaders and the patenting controversy over genes and gene products. Such a connection might not have been made had it not been for the May 18, 1995 press event. Now that it has, however, it is worthy of some theological investigation and ethical discernment. This is what I have tried to offer here. Let me conclude with some summary observations, actually recommendations based on the discussion above. Here is what I recommend we do.

1. Disregard the attempt to draw a connection between divine creation and a ban on patents as proposed by the Joint Appeal Against Human and Animal Patenting. The theology enunciated there is at best vague if not downright incoherent; and it fails even to articulate the fundamental commitments of the Hebrew and Christian traditions. There is no warrant for treating nature prior to human technological intervention as the sole domain of divine creation, let alone as more sacred than nature after it has been influenced by human ingenuity.

2. Pursue rather a theological vision that affirms continuity between humanity and the rest of nature, a vision that affirms our divine call to be stewards of our opportunities to strive toward relieving suffering and making the world a better place. Pursue a dynamic understanding of continuing creation that includes human responsibility within it.

3. Avoid superficial and misleading rhetoric that confuses and inflames the already existing controversy regarding the patenting of DNA knowledge and transgenic life forms. To speak of patenting human body parts, embryos, and fetal parts is wild speculation that has no connection to what is actually happening. In addition, we should ask for increased precision regarding the term *gene patent*.

If the word gene here refers to what exists naturally in the human genome, then no patent should be issued. If *gene* refers to a cloned sequence in a cDNA or EST that matches the original genome, then likewise no patent should be issued. If, however, *gene* refers to an altered or engineered DNA sequence that is novel, useful, and nonobvious, then intellectual property protection should be considered.

4. Take seriously the concern expressed by the Joint Appeal Against Human and Animal Patenting that the dignity and integrity of life be safeguarded. Noting that there is an enormous difference between a patented cell line and a living human being, we need to remain alert lest we reduce the latter to the former. Ever present is the temptation to "thingify" or commodify and commercialize the precious subjective dimensions of sentient life. We should provide guidelines so that the actual animals and actual persons whose bodies bear genetically modified cells retain the dignity and integrity they deserve.

5. Support a public policy that maintains the distinction between *discovery* and *invention*, between learning what already exists in nature and what human ingenuity creates. The intricacies of nature, no matter how fascinating to the one who uncovers them, ought not to be patentable. This applies to raw knowledge of the existing human genome. Preserve the Douglas principle. Novel creations through genetic engineering that are likely to contribute to better health for the human race should be dubbed useful, and therefore eligible for patent consideration.

6. Minimize government patenting. Federally funded research should aim primarily to enrich the public domain of knowledge. If James Watson and Francis Collins and other working scientists anticipate that patents will obstruct the free flow of scientific knowledge, then open up the free flow. If the biotech trade groups feel that government patenting is unnecessary and perhaps even burdensome, and if these represent the very sector the patenting policy was intended to serve, then the reason to proceed with this policy has been removed. Remove the policy. Support the NCHGR directive to discourage patent applications on raw genomic data. Or, better, provide legislation to permit the PTO to turn down such applications.

7. Foster an egalitarian philosophy for government funded medical research. Although it is prudent for federal agencies to fund research that will eventually stimulate new products in the private sector, selection principles are called for. Because some genetically

linked diseases such as Alzheimer's afflict a large number of people and hence create a large and potentially profitable market, other less widespread conditions such as Williams Syndrome involve relatively few in number, and offer smaller markets and lower levels of profitability. Venture capital will naturally flow in large amounts toward research on Alzheimer's where the expected dollar return is high. Those genetically linked diseases with lower frequency could be disregarded by the private sector. It would be an egalitarian move on the part of the NIH to survey repeatedly the broad scope of genetic research and support those areas overlooked by the strictly for-profit enterprises.

8. Treat all living animals, even transgenic animals, with a level of dignity that lifts them up above mere inanimate things, taking maximum care to minimize suffering to the extent possible in each situation. We should not allow ourselves to presume that just because an animal has been engineered genetically that it is akin to an engineered mechanical device. Animals are not mere things, even if their genetic make-up is the result of a technological process.

< SIX >

The Question of Germline Intervention

And genetic power is far more potent than atomic power. And it will be in everyone's hands. It will be in kits for backyard gardeners. Experiments for schoolchildren. Cheap labs for terrorists and dicta- tors. And that will force everyone to ask the same question—What should I do with my power?—which is the very question science says it cannot answer.

—Michael Crichton, *Jurassic Park*

Redemption fulfills creation; it does not transform it into something else or even something "higher."

—Langdon Gilkey

While wrestling with the interaction of the material and spiritual dimen- sions of human nature, the midcentury Roman Catholic theologian Karl Rahner described the evolutionary history of the human race in terms of "becoming." Human becoming consists in the self-transcendence of living matter. Nature has a history, and this history develops toward the human experience of freedom in the spirit. But it does not stop there. Nature will progress through and beyond the human stage toward the consummation of the cosmos as a whole, a consummate fulfillment yet to be achieved despite—yet through—the free human spirit. The human race is not merely a spiritual observer of material nature. Nor is human history lim- ited to cultural history. Rather, says Rahner, human history is "also an active alteration of this material world itself."[1] We human beings apply our "technical, planning power of transformation" even to ourselves. As subject we are becoming our own object, becoming our own creator.[2]

Curiously, what Rahner is describing here as human nature is feared by many as usurping divine nature. "Playing God" is the phrase invoked by many to shout "No!" to the attempt by the human race to influence its own evolution.

The acerbic rhetoric that usually employs the phrase "playing God" is aimed at inhibiting if not shutting down certain forms of scientific research and medical therapy. This applies particularly to the field of human genetics and, still more particularly, to the prospect of germline intervention especially for purposes of human enhancement—that is, the insertion of new gene segments of DNA into sperm or eggs before fertilization or into undifferentiated cells of an early embryo that will be passed on to future generations and may become part of the permanent gene pool.[3] Some scientists and religious spokespersons are trying to shut the door to germline intervention and tack up a sign reading "Thou shalt not play God."

Our task here will be to show how the proscription against playing God is being applied to arguments regarding germline alteration, especially the arguments raised by the Council for Responsible Genetics in its "Position Paper on Human Germ Line Manipulation." We will see that much of this discussion is thoughtful and wholesome. The ethicists of our day should be congratulated for engaging in pioneering work. I will, however, take the opportunity in this analysis to render a critique of some of the arguments raised against germline intervention. Although I recognize with others that great caution must be taken, I do not believe the dangers call for a lack of vision or a lack of courage. The theological concept of anthropology with which we have been working emphasizes human creativity placed in the service of visionary beneficence, and I think that even germline modification should be considered one possible means oriented toward a beneficent end. I have been arguing that if we understand God's creative activity as giving the world a future, and if we understand the human being as a created cocreator, then ethics begins with envisioning a better future. This suggests we should at minimum keep the door open to improving the human genetic lot and, in an extremely modest way, influencing our evolutionary future.[4] The derisive use of the phrase, "playing God," should not deter us from shouldering our responsibility for the future. To seek a better future is to "play human" as God intends us to.[5]

Somatic vs. Germline; Therapy vs. Enhancement

These issues come to the forefront of discussion due in large part to the enormous impact of the Human Genome Project on the biological and even the social sciences. Descriptively, we know the stated purposes directing the Human Genome Project as presently conceived. First, its aim is knowledge. The simple goal that drives all pure science is present here, namely, the desire to know. In this case it is the desire to know the sequence of the base pairs and the location of the genes in the human genome. Second, its aim is better human health. The avowed ethical goal is to employ the newly acquired knowledge from research to provide therapy for the many genetically caused diseases that plague the human family. John C. Fletcher and W. French Anderson put it eloquently: "Human gene therapy is a symbol of hope in a vast sea of human suffering due to heredity."[6] As this second health-oriented purpose is pursued, the technology for manipulating genes will be developed, and questions regarding human creativity will arise. How should this creativity be directed?

Virtually no one contests the principle that new genetic knowledge should be used to improve human health and relieve suffering. Yet a serious debate has arisen that distinguishes sharply between therapy for suffering persons who already exist and the health of future persons who do not yet exist. It is the debate between somatic therapy and germline intervention. By somatic therapy we refer to the treatment of a disease in the body cells of a living individual by trying to repair an existing defect. It consists of inserting new segments of DNA into already differentiated cells such as those that we find in the liver, muscles, or blood. Clinical trials are underway to use somatic modification as therapy for people suffering from diabetes, hypertension, and Adenosine Deaminase Deficiency. By germline therapy, however, we refer to intervention into the germ cells that would influence heredity and hopefully improve the quality of life for future generations. Negatively, germline intervention might help to eliminate deleterious genes that dispose us to disease. Positively, though presently well beyond our technical capacity, such intervention should certainly actually enhance human health, intelligence, and strength.

Two issues overlap here and we should sort them out for clarity. One is the issue of somatic intervention versus germline intervention. The other is the issue of therapy versus enhancement. Although somatic treatment is usually identified with therapy and germline treatment with enhancement, there are occasions where somatic treatment enhances, such as injecting growth hormones to enhance height for playing basketball. And germline intervention, at least in its initial stages of development, will aim at preventive medicine. The science of enhancement, if it comes at all, will only come later.

Every ethical interpreter I have reviewed agrees that somatic therapy is morally desirable and looks forward to the advances gene research will bring for expanding this important medical work. Yet many who reflect on the ethical implications of the new genetic research stop short of endorsing genetic selection and manipulation for the purposes of improving the human species.[7] The World Council of Churches (WCC) is representative. In a 1982 document, we find

> somatic cell therapy may provide a good; however, other issues are raised if it also brings about a change in germline cells. The introduction of genes into the germline is a permanent alteration. . . . Nonetheless, changes in genes that avoid the occurrence of disease are not necessarily made illicit merely because those changes also alter the genetic inheritance of future generations. . . . There is no absolute distinction between eliminating "defects" and "improving" heredity.[8]

The text elsewhere indicates that the WCC is primarily concerned with our lack of knowledge regarding the possible consequences of altering the human germline. The problem is this: the present generation lacks sufficient information regarding the long-term consequences of a decision today that might turn out to be irreversible tomorrow. Thus, the WCC does not forbid forever germline therapy or even enhancement. Rather, it cautions us to wait and see. In a similar fashion, the Methodists "support human gene therapies that produce changes that cannot be passed on to offspring (somatic), but believe that they should be limited to the alleviation of suffering caused by disease."[9] The United Church of Christ also approves "altering cells in the human body, if the alteration is not passed to offspring."[10] On June 8, 1983 fifty-eight religious leaders issued a "Theological Letter Concerning the Moral Arguments" against germline engineering addressed to the U.S. Congress. The group action was orchestrated by Jeremy Rifkin of the Foundation on Economic Trends. One member, James R. Crumley, presiding bishop of the then Lutheran Church in America spoke to the press saying, "There are some aspects of genetic therapy [for human diseases] that I would not want to rule out. . . . My concern is that someone would decide what is the most correct human being and begin to engineer the germline with that goal in mind."[11]

A more positive approach is taken by The Catholic Health Association. If we can improve human health through germline intervention, then it is morally desirable.

> Germline intervention is potentially the only means of treating genetic diseases that do their damage early in embryonic development, for which somatic cell therapy would be ineffective. Although still a long way off, developments in molecular genetics

suggest that this is a goal toward which biomedicine could reasonably devote its efforts.[12]

Part of the reluctance to embrace germline intervention has to do with its implicit association with the history of eugenics. The term eugenics brings to mind the repugnant racial policies of Nazism, and this accounts for much of today's mistrust of genetic science in Germany and elsewhere.[13] No one expects a repeat of Nazi terror to emerge from genetic engineering; yet some critics fear a subtle form of eugenics may be slipping in the cultural back door.[14] John Harris may be a bit of a maverick, but he welcomes eugenics if it contributes to better human health. He makes the point forcefully: "where gene therapy will effect improvements to human beings or to human nature that provide protections from harm or the protection of life itself in the form of increases in life expectancy . . . then call it what you will, eugenics or not, we ought to be in favor of it."[15]

Philosophical and ethical objections to eugenics seem to presuppose not therapy but rather enhancement. The growing power to control the human genetic make-up could foster the emergence of the image of the "perfect child" or a "super strain" of humanity. Some religious leaders worry that the impact of the social value of perfection will begin to oppress all those who fall short. Ethicists at the March 1992 conference on "Genetics, Religion and Ethics" said this:

> Because the Jewish and Christian religious world-view is grounded in the equality and dignity of individual persons, genetic diversity is respected. Any move to eliminate or reduce human diversity in the interest of eugenics or creating a "super strain" of human being will meet with resistance.[16]

In sum, with the possible exception of the Catholic Health Association, religious ethical thinking tends to be conservative in the sense that it seeks to conserve the present pool of genes on the human genome for the indefinite future.

Now the question of playing God begins to take on the form of the *Frankenstein* or *Jurassic Park* fever. The risk of exerting human creativity through germline intervention is that, though we begin with the best of intentions, the result may include negative repercussions that escape our control. Physically, our genetic engineering may disturb the strength-giving qualities of biodiversity that we presume contributes to human health. Due to our inability to see the whole range of interconnected factors, we may inadvertently disturb some sort of existing balance in nature and this disturbance could redound deleteriously. Socially, we could contribute to stigma and discrimination. The very criteria to determine just what counts as a "defective" gene may lead to stigmatizing all those persons who carry that gene. The very proffering of the image of the ideal child or a super strain of humanity may cultivate a sense of inferiority to those who do not measure up. To embark on a large scale program of germline enhancement may create physical and social problems, and then we would blame the human race for its pride, its *hubris*, its stepping beyond its alleged God-defined limits that brings disaster upon itself.

Yet, there may be another way to look at the challenge that confronts us here. The correlate concepts of God as the creator and the human as the created cocreator orient us toward the future, a future that should be better than the past or present. One of the problems with the naturalist argument and the more conservative religious arguments mentioned above is that they implicitly assume the present state of affairs is adequate. These arguments tacitly bless the *status quo*. The problem with the *status quo* is that it is filled with human misery, some of which is genetically caused. It is possible for us to envision a better future, a future in which individuals would not have to suffer the consequences of genes such as those for Cystic Fibrosis, Alzheimer's or Huntington's Disease. That we should be cautious and prudent and recognize the threat of human *hubris*, I fully grant. Yet, our ethical vision cannot acquiesce with present reality; it must press on to a still better future and employ human creativity with its accompanying genetic technology to move us in that direction.

Germline Enhancement: A Closer Look

Having enunciated my own conviction that the Christian doctrine of creation with its accompanying understanding of the human being as the created cocreator leads to an ethic oriented toward striving for a better future, let us turn to a closer look at the arguments for and against germline intervention and manipulation. Eric T. Juengst helpfully summarizes five arguments in favor of germline modification for the purposes of therapy.

(1) *Medical utility*: germline gene therapy offers a true cure for many genetic diseases.
(2) *Medical necessity*: such therapy is the only effective way to address some diseases.
(3) *Prophylactic efficiency*: prevention is less costly and less risky than cure.

(4) *Respect for parental autonomy* when parents request germline intervention.
(5) *Scientific freedom* to engage in germline inquiry.

Juengst also summarizes five arguments opposing germline intervention.

(1) *Scientific uncertainty and risks* to future generations.
(2) *Slippery slope to enhancement* that could exacerbate social discrimination.
(3) *Consent of future generations* is impossible to get.
(4) *Allocation of resources*: germline therapy may never be cost effective.
(5) *Integrity of genetic patrimony*: future generations have the right to inherit a genetic endowment that has not been intentionally modified.[17]

In order to engage the issue in some detail and to test a theological commitment to the notion of the created cocreator, I would like to turn our attention to a representative case in point, namely, the position paper drafted by the Council for Responsible Genetics (CRG) in the fall of 1992. The CRG proffers three types of argument in opposition to germline modification in humans: a technical argument, a slanderous argument, and an ethical argument.

The first argument against germline manipulation is technical. Although the motive for modifying germ genes may be the enhancement of human well-being for future generations, unexpected deleterious consequences may result. Removal of an unwanted disease gene may not eliminate the possibility that other gene combinations will be created that will be harmful. Inadvertent damage could result from biologists' inability to predict just how genes or their products interact with one another and with the environment. "Inserting new segments of DNA into the germline could have major, unpredictable consequences for both the individual and the future of the species that include the introduction of susceptibilities to cancer and other disease into the human gene pool."

It would seem to the prudent observer that we take a "wait and see" attitude, that we move cautiously as the technology develops. We saw this in the WCC position. However, the problem of unexpected consequences is one that confronts all long-term planning, and in itself should not deter research and experimentation guided by a vision of a healthier humanity.[18]

The second argument appeals to guilt by association and is thereby slanderous. The CRG Human Genetics Committee says "the doctrine of social advancement through biological perfectibility underlying the new eugenics is almost indistinguishable from the older version so avidly embraced by the Nazis." The structure of this argument is that because germline modification can be associated with eugenics, and because eugenics can be associated with Nazism, it follows that we can associate proponents of germline enhancement with the Nazis and, on this ground, should reject it. The argument borders on the *ad hominem* (circumstantial) fallacy.

One problem is that the CRG argument is too glib, failing to discern the complexities here. The eugenics movement of the late-nineteenth and early-twentieth centuries was originally a socially progressive movement that embraced the ideals of a better society. In England and America it became tied to ethnocentrism and the blindness of class interests, leading to forced sterilization of feeble-minded prisoners. It was eventually discarded because advances in genetics proved it unscientific.[19] In Germany the eugenics movement became tied to anti-Semitism, resulting in the racial hygiene (*Rassenhygiene*) program of the Nazi SS and the atrocities of the so-called "final solution."[20] With this history in mind, the present generation must certainly be on guard against future programs of "ethnic hygiene" which seem to plague the human species in one form or another every century. Yet we must observe that ethnocentric bias in England and America and the rise of Nazism in Germany were social phenomena that employed eugenics for their respective ends. Eugenics is not the source of injustice, even if it can be a weapon in the service of injustice. The CRG's use of the volatile word "Nazi" in this discussion of germline enhancement is an attempt to paint their opponents in such a repulsive color that no one will open-mindedly view the matter.

The third CRG argument, the ethical argument, is much more worthy of serious consideration. The central thesis here is that germline modification will reinforce existing social discrimination. The position paper declares,

> The cultural impact of treating humans as biologically perfectible artifacts would be entirely negative. People who fall short of some technically achievable ideal would increasingly be seen as "damaged goods." And it is clear that the standards for what is genetically desirable will be those of the society's economically and politically dominant groups. This will only reinforce prejudices and discrimination in a society where they already exist.[21]

Let us look at this argument in terms of its component parts. The assumption in the first sentence is that germline intervention implies biological perfectibility and, on account of this, that human persons will be treated as artifacts. It is of course plausible that a social construction of the perfect child or the perfected human strain might appear in Saturday morning cartoons and other cultural forms. Yet, this does not seem to apply to the actual situation in which genetic scientists currently find themselves. They are occupied with much more modest aspirations such as protection from monogenetic diseases such as Cystic Fibrosis. The medical technology here is not much beyond infancy. At this point in technological history we do not find ourselves on the brink of designer children or the advent of a super strain. What is "genetically desirable" is by no means scientifically attainable. Thus, Hessel Bouma and his colleagues are less worried than the CRG because they recognize that the technological possibility of creating a genetically perfect human race is still very remote. "Things like intelligence and strength are not inherited through single genes but through multifactoral conditions, combinations of inherited genes and numerous environmental factors. Our ability to control and to design is limited by the complexity of many traits, so there are seemingly insurmountable technological and economic barriers that weaken the empirical slippery-slope argument that we are sliding into the genetic engineering of our children."[22]

Continuing our analysis of the ethical argument, the CRG rightly alerts us to the social-psychology of feeling like "damaged goods" and being treated like "damaged goods." If a "technically achievable ideal" should become a cultural norm, then those who fail to meet the norm would understandably feel inferior.[23] Furthermore, the economically and politically advantaged groups will help to steer the definition of the ideal norm to serve their own class interests.[24] Here the CRG should be applauded for alerting us to a possible loss of human dignity.

At this point a reaffirmation of human dignity is called for, I believe, wherein each individual person is treated as having the full complement of rights regardless of his or her genes. Ethical support here comes from the Christian doctrine of creation, wherein God makes men and women in the divine image and pronounces them "good" (Genesis 1:26–31). It also comes from the ministry of Jesus, wherein the Son of God sought out the outcasts, the lame, the infirm, the possessed—surely those who were considered the "damaged goods" of first-century Palestine—for divine favor and healing.[25] Each human being, regardless of health or social location or genetic endowment is loved by God, and this recognition should translate into social equality and mutual appreciation. There is no theological justification for thinking of some persons as inferior to others, and new technical possibilities in genetics ought not change this.

We also note the CRG's prognostication for the future: germline modification "will only reinforce prejudices and discrimination in a society where they already exist." Prejudices and discrimination exist in the present, says the CRG. This is an obvious fact we readily concede. Does it follow, however, that germline intervention "will only reinforce" them? Is germline modification the cause of present prejudice and discrimination? No. Prejudice and discrimination seem to flourish quite well without germline manipulation, yet somehow this is alleged to count as an argument against the latter.

If the argument rests on the premise that germline enhancement will create a technical ideal achievable by some but not others, then it fails on the grounds of triviality. This could apply to countless ideals in our society. We daily confront innumerable ideals that are met by some but not all, whether they be athletic achievements, beauty trophies, professional promotions, or lottery winnings. These may elicit temporary feelings of inferiority on the part of those who come in second or further behind, but they are widely ignored by those who did not compete. Given the realistic prospects for what germline enhancement is aimed at accomplishing, the new situation would not alter the present situation in this respect. If it is technically possible to relieve some individuals from suffering the consequences of diabetes through the regular use of insulin, then the achievement of this ideal for those afflicted by diabetes leads to only gratitude on their part and on the part of those who love them. Somatic cell therapy or even germline modification for diabetes will only extend this gratitude. To those who are not afflicted or likely to be afflicted by diabetes, this achievement may be applauded from a distance or perhaps ignored.

One could envision a next step, of course, where germline intervention could, if made universally available, eliminate the likes of diabetes from the human gene pool. We would then have a future wiped clean of genetically based diabetes. If this constituted an achieved ideal for the whole human race, and if the unexpected consequences were less harmful than the diabetes, then many persons will have been spared the suffering diabetes *could* have caused and no reinforcement of prejudice and discrimination will have occurred.

What if we were to falter somewhere along the way? Suppose we began a worldwide program to eliminate the predisposition to diabetes from the human gene pool, achieved success in some family or ethnic or class groups, and then due to lack of funding or other factors had to abandon the project. What would happen to those individuals who still carried the deleterious gene? Would they suffer stigma or discrimination? Perhaps, yes. And the CRG rightly alerts us to such a possibility. Yet, we might ask, does this prospect provide sufficient warrant to shut down the research and prohibit embarking on such a plan?

Inter-Generational Genethics

The CRG buttresses the central ethical argument with two subarguments. One is that the present generation, presumably the one engaging in germline modification, cannot be held accountable by future generations for the wrongful damage we inflict on them. We, our progeny's ancestors, will not be around any more to be accountable. There may be an equivocation at work here. On the one hand, the present generation will be absent in the future and, therefore, we cannot be held accountable in the sense that we can be punished by imprisonment. On the other hand, though absent, we can be held accountable in the sense that future fingers could be pointed and fists thrown into the air as our progeny express anger at our failure to assume responsibility. Just because we cannot be punished does not mean we are not accountable in a moral sense.

Yet, for the CRG, somehow the concept of accountability is supposed to count against germline enhancement. Again the argument fails on account of triviality, because our responsibility to our progeny applies across the board to all departments of life. There is nothing special about genes. One might even make a case that environmental responsibility is of graver ethical concern. The excessive depletion of nonrenewable natural resources and pollution of the biosphere is due to the hedonism of the present generation, due to present selfishness that sacrifices the welfare of future generations for the prosperity of our own. Germline intervention, in contrast, could be motivated only by seeking benefit for future generations whom we may not live to see. With or without accountability, the latter at least has the virtue of altruism going for it.

The other subargument raises an interesting issue worth pondering further. The CRG says: "Germline modification is not needed in order to save the lives or alleviate suffering of existing people. Its target population are 'future people' who have not yet even been conceived." On the face of it, this argument looks like another brand of defense for the ecological hedonists just mentioned, whose interests are limited to only the present generation without any regard for future progeny. But this may be a misreading. The CRG is not eliminating our responsibility for future generations. Yet, for some unexplained reason, the CRG makes central the distinction between people who exist and people who do not yet exist. The assumption is that moral priority is given to those who exist over against those who have "not yet even been conceived." The interesting puzzle is the relative moral status of present and future, existents and not yets.

Suppose we draw up the previous concern for accountability and combine it with the concepts of rights and wrongful birth. Might future generations blame us today for their wrongful birth by damaging them through germline intervention? Or, in contrast, might they blame us for not intervening in the germline, thereby leaving them to suffer from diseases we could have prevented? We are at an ethical crisis—that is, on the verge of an ethical challenge where creative action is demanded—because whether we engage in germline intervention or not, if we are technically capable, then we will be morally accountable.

Here the contrast with the environmental crisis is illuminating. We can imagine our great grandchildren raising their fists in anger at us, living on a deforested earth, mines depleted of their minerals, lakes dead from acid rain, food supply contaminated by chemicals, skin cancerous due to depleted ozone layer. They will claim we violated their right to a life-giving environment and, despite what the CRG says, they will claim we are accountable as they burn us in effigy.

Does this apply by analogy to germline enhancement? We can certainly imagine a future person asserting, "My parents, grandparents, and great-grandparents and the genetic scientists of their generation violated my rights by giving me a bad genetic endowment." It would be a variant on the wrongful-birth accusation. Yet, not everyone sees the sense this makes. Hardy Jones, for example, would argue: had this individual's progenitors taken successful steps toward enhancing the genetic endowment of their offspring, then this would not be the child they actually had. Having a child with defective genes cannot be a violation of that child's right, because it is not possible to respect that right by not having the child or by bequeathing a different genetic constitution. The only child who can claim a right is one that exists, and the particular configuration of genes is definitional to the person who exists. "Genetically defective persons are not analogous to existing individuals who subsequently acquire biologically bad qualities."[26]

Reproductive liberty advocate John A. Robertson makes a similar argument when asking about the consent or lack of consent on the part of future generations to what we do today to affect their germline. If no harm occurs, he argues, then this is a mere theoretical objection. If harm does occur, then the question of identity arises. "Later generations allegedly harmed without their consent may not have existed at all. Different individuals would then exist than if the germline gene therapy had not occurred."[27]

Perhaps the CRG position-paper writers presumed this kind of distinction between existing and not yet existing persons, and this permitted them to give qualified approval of somatic modification for living persons while proscribing germline manipulation.[28] What this means for us, then, is that if we are to affirm ethical responsibility for the genetic inheritance we bequeath our progeny, then the framework of rights and accountability might be inadequate. As long as the CRG works within this framework, perhaps its conclusions are understandable.

The Not-Yet-Future and the Ethics of Creativity

Would a future-oriented theology of creation and its concomitant understanding of the human being as God's created cocreator be more adequate? It would be more adequate for a number of reasons. First, a future-oriented theology of creation is not stymied by giving priority to existing persons over future persons who do not yet exist. A theology of continuing creation looks forward to the new, to those who are yet to come into existence as part of the moral community to which we belong. Second, such a theology is realistic about the dynamic nature of our situation. Everything changes. There is no standing still. What we do affects and is affected by the future with its array of possibilities. We are condemned to be creative for good or ill. Third, the future is built into this ethical vision. Once we apprehend that God intends a future, our task is to discern as best we can the direction of divine purpose and employ that as an ethical guide. When we invoke the apocalyptic symbol of the New Jerusalem where "crying and pain will be no more" (Revelation 21:4), this will inspire and guide the decisions we make today that will affect our progeny tomorrow.

The creative component to a future-oriented ethic denies that the status quo defines what is good, denies that the present situation has an automatic moral claim to perpetuity. Take social equality as a relevant case in point. As one can plainly see, social equality does not at present exist, nor has it ever existed in universal form. We daily confront the frustrations of economic inequality and political oppression right along with the more subtle forms of prejudice and discrimination that the CRG rightly opposes. Human equality, then, is something we are striving for, something that does not yet exist but ought to exist. Equality needs to be created, and it will take human creativity under divine guidance to establish it plus vigilance to maintain it when and where it has been achieved. Wolfhart Pannenberg, who has developed an ontology of the future, puts it this way: "The Christian concept of equality does not mean that everyone is to be reduced to an average where every voice is equal to every other, but equality in the Christian sense means that everyone should be raised up through participation in the highest human possibilities. Such equality must always be created; it is not already there."[29] An ethic that seeks to raise us to the "highest human possibilities" cannot accept the *status quo* as normative, but presses on creatively toward a new and better future. Applied to the issue at hand, Ronald Cole-Turner makes the bold affirmation: "I argue that genetic engineering opens new possibilities for the future of God's creative work."[30]

Conclusion

We began this chapter with an observation of Karl Rahner regarding evolution and human openness toward the future. Self-transcendence and the possibility for something new belong indelibly to human nature. Human existence is "open and undetermined."[31] That to which we are open is the infinite horizon; we are open to a fulfillment yet to be determined by "the infinite and the ineffable mystery" of God.[32] If we try to draw any axioms that connect this sublime theological vision to an ethic appropriate to genetic engineering, then openness to the future translates into responsibility for the future—even our evolutionary future. Such a theological vision undercuts a conservative or reactionary proscription against intervening in the evolutionary process. Rahner describes the temptation to condemn genetic research and its application as "symptomatic of a cowardly and comfortable conservatism hiding behind misunderstood Christian ideals."[33] The concept of the created cocreator we invoke here is a cautious but creative Christian concept that begins with a vision of openness to God's future and responsibility for the human future.

The health and well-being of future generations not yet born is a matter of ethical concern when viewed within the scope of a theology of creation that emphasizes God's ongoing creative work and that pictures the human being as the created cocreator. A vision of future possibilities, not the present *status quo*, orients and directs ethical activity. When applied to the issue of germline intervention for the purpose of enhancing the quality of human life, the door must be kept open so that we can look through, squint, and focus our eyes to see just what possibilities loom before us. This will include a realistic review of the limits and risks of genetic technology.[34] But realism about technological limits and risks is insufficient warrant for prematurely shutting the door to possibilities for an improved human future. Rather than playing God or taking God's place, seeking to actualize new possibilities means we are being truly human.

< SEVEN >

A Theology
of Freedom

Sound the Timbrel o'er Egypt's dark sea! Jehovah has triumphed—his
people are free.

—Thomas Moore, 1779–1852

In the previous chapters of this book we have engaged genetic science and
the gene myth with response theology. We have employed a hermeneutic
of secular experience. By examining the direction in which scientific
research is taking us and by analyzing the implications of the gene myth,
we have identified points of contact where theological response is called
for. My hope is that by responding with theological reflection on these scientific and cultural forms of understanding, I have been able to illuminate
some aspects of our current situation.

Here in this final chapter we shall turn more directly to theological
resources for apprehending what is at stake in understanding human freedom. Of the four freedoms we have been working with, moral freedom
and future freedom will receive primary attention, although political liberty and free will will not disappear from the horizon of discussion.

The field within which we are working here is systematic theology.
"The systematic theologian's major task is the reinterpretation of the tradition for the present situation," writes David Tracy at the University of
Chicago.[1] To pursue this task the systematic theologian asks what I call
the "hermeneutical question," the question that defines modern theology
as modern: *How can the Christian faith, first experienced and symboli-
cally articulated in an ancient culture now long out of date, speak meaningfully to human existence today as we experience it amid a worldview*

dominated by natural science, secular self-understanding, and the world-wide cry for freedom? As the modern world teeters on the edge of post-modernity with its unresolved battles between pluralism and unity, I add an amendment that takes the form of a related question: *How can the Christian faith be made intelligible amid an emerging postmodern consciousness that, although driven by a thirst for both individual and cosmic wholeness, still affirms and extends such modern themes as evolutionary progress, future consciousness, and individual freedom?*

The structure of these questions compels the systematic theologian to engage in a theology of culture. A close reading of culture, even very secular aspects of culture such as natural science, reveals that at its depth culture is spiritual. In the depth of culture we find religious questions and the search for spiritual meaning. This is an insight developed by existentialist theologian Paul Tillich for whom "Theology of Culture" aims at uncovering the deeper religious questions lying below the surface of cultural forms. The theologian does not need to bring the formal doctrines of his or her specific tradition or institutionalized religion in order to deal with religious meanings. Spiritual insights and quests are already present in culture; in fact, they animate culture. "Culture is the form of religion and religion the substance of culture," Tillich was known for saying.[2]

The appropriation of Tillich's insight for examining theologically the specific role of natural science has been furthered by Langdon Gilkey, a Tillich disciple. For Gilkey, the theology of culture seeks "to understand what is happening in contemporary culture ... it asks questions about the *religious* dilemmas of cultural life and so questions guided by *theological* problems and concerns. By such religious or theological questions I mean questions of meaning and meaninglessness, of the ambiguity even of creativity, of the freedom *and* the bondage of the will, of the strange, inexplicable inheritance of evil, of the career of good and evil in the passage of time, of the contradiction of even our highest values, of the tension between affirmation and tolerance, between pluralism and truth, and of the promise of new possibilities and the need for hope for the future."[3] These are the questions we have been asking as we examined the advance of genetic science and its cultural translation into the gene myth.

Theologians too give expression to the culture that surrounds us and works through us. We systematic theologians have our own social location within which we find ourselves. We are products of our culture as well as producers of cultural analysis. Cultural analysis is self-analysis. This means that questions regarding determinism and freedom are internal to theology. We do not need to wait for the sciences to pose them. In fact, many of the contemporary formulations of issues regarding freedom in post-Enlightenment culture are secular reiterations of previous theological agendas. So, to employ distinctively theological resources in pursuing the question of freedom, with its concomitant concerns for human selfhood and human personhood, is itself a cultural enterprise.

A focal example of a previous theological issue that has garnered a modern secular counterpart is the ancient question: Does God's predestination obviate human freedom? The modern secular version is this: Does determinism in nature obviate human freedom? Both versions seem to presuppose the mistaken belief that determination—whether divine determination or natural determination—is exclusive of human freedom. However, these two are not incompatible, either theologically or scientifically.

The theological mistake begins with what I call the fixed-pie assumption. Unfortunately, theologians sometimes assume that there is a fixed amount of power in the universe. So, if God has more power, then God's creatures must have less. If God is omnipotent, then we creatures have no power at all. If God is all powerful, then we must be determined, totally determined. But this, in my judgment, commits the quantification fallacy. It assumes that God and human beings are in competition over a finite amount of power. However, there is no fixed power-pie. God is dynamically and qualitatively related to the world. The divine dynamism does not require stealing power from the universe or withholding power from creation in order for God to remain omnipotent. Quite the contrary: God's creative and redeeming work imparts power to the world. When God's Holy Spirit enters the believer he or she feels empowered. God's power generates selves who become loci of power and, hence, free. The exertion of divine power effects liberation for us. God's freedom breeds human freedom.

Similarly, genetic determinism could not be in competition with human freedom. There is no fixed pie of determinism, so that if our DNA gets a bigger slice we get a smaller quantity of self-determining freedom. Our individual genomes contribute one set of determinants among others that constitute the overall conditioned character of our life. The "conditionedness" of our life simply represents the finite point of departure for the expression of human freedom. What we mean by freedom is the expression of a self, a human person. Gene expression and human expression are not the same thing. Genes do not make decisions. Human beings do. Human beings are whole organisms of which the genes are necessary and vital parts, to be sure; but no human person can be reduced to the mere sum of his or her constituting parts. The whole self is greater than the sum of the parts that make us up.

The opposite of determinism is indeterminism, *not freedom*. Indeterminism consists in random chance. Freedom begins with deliberation, decision, and responsible action on the part of human persons. The alleged conflict between genetic determinism and human freedom is a category mistake. We can be both determined and free at the same time.

Claremont theologian Marjorie Hewitt Suchocki says that "freedom is always conditioned. . . . Indeterminism is a blind responsiveness to influences; freedom is the ability to lift this responsiveness to consciousness."[4] The highest form of freedom, according to Suchocki, has to do with the development of personhood. It is what some might think of as self-actualization. It is the "freedom to develop according to one's highest potential."[5] Yet this "freedom as the ability to realize one's potential rests upon a more fundamental definition of freedom as the ability to choose."[6] Persons choose, and by choosing they can further the development of their own selfhood. The power to choose is one of the gifts from God bestowed in behalf of our liberation.

In what follows we shall ask and reask a number of questions already addressed regarding human freedom. This time we shall do so in conversation with a number of theologians who have given serious thought to this issue. Does human freedom require that we liberate ourselves from nature, even our own nature in the case of DNA determinism? We will conclude: no, human freedom does *not* require liberation from nature. Nature, including our genetic makeup, establishes the particular conditions for the particular ways in which we as persons will exercise our freedom.

Where do we find the human self, the human person, that takes the initiative we know as exercising free will? The location is the metaxy, the tensive relationship between soil and spirit. We human beings are soil in that we belong to the earth, to nature. By soil here I mean our physical make-up, including the DNA bequeathed to us by evolutionary history. Yet we are more than soil. We are spirit too. Our imaginations bear our thoughts beyond earthbound limits in the direction of infinite possibility. The metaxy is the creative tension between past limits and future possibilities. What I mean by the human soul is inclusive of soil and spirit. The soul plays host to the metaxic tension.

I will add that we are in the process of becoming human. We are not there yet. In addition to our evolutionary past we need to factor in our divinely appointed destiny, a destiny that includes resurrection from the dead.

Does divine freedom and power enhance or compete with human freedom and power? Clearly, from the theological perspective, God's freedom is the source of human freedom. It is God's call forward to a creative future that draws human consciousness beyond soil to spirit. No competition exists between divine power and human freedom, because God exercises divine power on behalf of human liberation.

If God exercises divine power to open the future for the exercise of human freedom, what form does responsible freedom take? Proleptic ethics—that is, we work creatively in the present in light of a projected vision of a redeemed future. Should we play God? No, we should not play God in the promethean sense. But we should play human in the *imago dei* sense—that is, we should understand ourselves as created cocreators and press our scientific and technological creativity into the service of neighbor love, of beneficence.

Do We Want Liberation from Nature?

In an earlier chapter we distinguished political freedom from natural freedom—that is, external or circumstantial freedom known as political liberty from freedom of the will, which we have classically assumed is given us by nature. To the extent that the puppet determinism of the gene myth holds water, nature is coming to look like a political power that restricts our freedom. What makes the threat of genetic determinism so fascinating and so ominous is that the genes are internal, not external, to us. DNA is nature on the inside.

External freedom is defined negatively. Political freedom or liberty, for example, is gained by negating the limits or constraints placed upon us by a dictatorial government. External freedom is gained through the process of liberation, the process of breaking the chains or cutting the bonds that tie us into submission. Now, let us ask: Is our relation to our internal nature like that of a subject to a political dictator? Does human freedom come with liberation from our DNA?

Existentialist theologian John Macquarrie leads us to ask this question when we see how he contrasts human nature from animal nature. Species of animals survive by adapting to their natural environment. Human beings have stood this evolutionary law on its head, so to speak, because we now survive by adapting our environment to ourselves. This is biological freedom. "The lion that is born free and stays free is not subject to any artificial or humanly contrived restraints," he observes, "but of course it is never for one moment delivered from the constraints of nature. Human freedom is far more fundamental. It is the negation of nature itself, a distancing from nature, a kind of declaration of human independence."[7] Macquarrie goes on to add that we cannot entirely free ourselves from nature's constraints, of course. We human beings are always partly free and partly determined by nature's laws. Yet, we need to ask: Is it necessary to understand human freedom as liberation from nature?

No. We human beings are fully natural. We belong fully to the soil, so to speak. DNA determinism, to the extent that it is influential, sets the initial conditions or parameters within which free will comes to particular expression. Our finite nature specifies us. It does not imprison us.

It is not from nature that we seek liberation. What we seek liberation from is the past, and we do so on behalf of an openness toward the future. Future freedom is an inherited gift of nature, even if it constitutes a dynamic principle within the history of nature that is itself tranformatory of nature. Macquarrie understands and appreciates this future dimension of human freedom. "It is often thought that freedom to choose between alternative courses of action is the fundamental form of freedom—'freedom of the will'—as it is usually called," he writes. "But it seems to me that this is too restrictive. The fundamental freedom is creativity, especially the human freedom to shape humanity itself."[8]

Nature does not stand still. It is on the move. So is the human race. Conscious and free, we exert a certain degree of cocreativity in this movement. Again, we need to think of the human being as the *created cocreator*.[9] Cocreativity illuminates the divine image in us, the *imago dei*. We have not created ourselves all by ourselves, of course. God has done that. To say that we are created beings is to acknowledge God as the creator. God creates out of nothing, *ex nihilo*. We do not. We cannot presume to play God in this sense. Yet, we can exert creativity within the ongoing processes of creation; and as creative beings we are called by God to exert our creativity in responsible fashion. Responsible creativity is playing human as God intends us to.

This version of Christian anthropology presupposes what we said earlier, namely, nature is good. What God creates is good. Human creative potential belongs to the natural world, to the goodness of creation. The problem we confront is this: what we have inherited from our evolutionary past is by no means an unmitigated good. Built right into our DNA history is a mixture of health and disease. Included in our bio-cultural history may be some predispositions to unacceptable behavior. Through science and through moral living we are able to go beyond what we have inherited, and to a limited degree at least we can guide the course of natural and historical events toward a better inheritance for the future. The goodness of nature should be understood as a dynamic goodness, as belonging to a history of nature in which the pursuit of the good is a divinely inspired process. Rather than see nature alone as the ontological source of the good, we need to see God as the source of the good bestowed in a redeeming and creative way. Rather than seek liberation from nature, the created cocreator seeks to be responsible within nature for the future of nature.

Person and World

What makes this responsibility possible is the existence of persons who can exercise a degree of freedom. Persons come in relationship: in relationship to themselves, to other selves, to the world around, to the world within, and to God.

Significant for our discussion here is the soul. Without giving it metaphysical status, I associate the soul with the self or the central locus of a person. It is the conscious, rational, discriminating, unifying, purposeful element in each of us that leads us in one direction rather than another. The soul or self tends to center the person and give directedness; and this makes it possible for us to speak of freedom rather than mere randomness. Persons exercise freedom only in relationship.

The self-world relationship is a blurry one, but it is key. On the one hand, the self is externally related to the world. The world is out there. The things in our world are objects that influence us from the outside or that we influence externally. On the other hand, we are internally related to the world. The world lies within. Biologically, the air we breathe and the food we eat come to constitute who we are in an ongoing dynamic process. Culturally, the language we speak and the ideas we adopt as our own are inherited and shared; and our thought processes are to some extent intersubjective and overlapping with those of others. Does DNA belong to our world?

Notice the term "world" here rather than "environment." We are refining to some extent the concept of environment as used by the environmental determinists whom we have mentioned in this book. By "world" we refer to that portion of the environment that is meaningfully related to us, that is present to us in the self-world relationship.

Tillich offers some help in clarifying the polar character of the self-world relationship. "A self is not a thing that may or may not exist," writes Tillich; "it is an original phenomenon which logically precedes all questions of existence."[10] As an original phenomenon, the self cannot exist at the end of a series of experiments or arguments that establish its existence. Rather, it sits right up front, at the beginning. It is presupposed. Only a thinking and acting self could pursue the question of whether there is a self or what the self might be. Rather than a thing that can be studied like other things or objectified like other objects, the human self is one term in the self-world relationship. Because relationship is involved, the self is both separated and connected to what is not the self, namely, the world. Tillich continues: "Being a self means being separated in some way from everything else, having everything else opposite one self, being able to look at it and to act upon it. At the same time, however, this self is aware that it belongs to that at which it looks."[11]

The strictly nonself over against which the self is situated is the environment. Our environment is not merely the sum total of all things that exist in the universe, regardless of whether or not we are related to them. Rather, the concept of environment is more specific. To be distinctively *our* environment, *we* must have some interaction with the things in it. The moon's surface lacking in atmosphere is one item in the universe, but it does not belong to my environment, except perhaps as a pleasant visual object on a clear night. The polluted air that comes over my house courtesy of the oil refineries in and near Richmond, California, however, does belong to my environment. I interact with it. Similarly, I live with all the other things in my environment. I have a self-other relationship with these things.

Telescoping, we might distinguish first the widest possible environment, namely, the universe as a whole. Within this we have *our* particular environment, which includes the physical and cultural stimuli with which we actually interact. Finally, we have the world—that is, our environment treated as a singularity. The idea of the world is a correlate concept to self, treating all that is in the environment as a unity in correlation to the unity of the self. The world is not the sum-total of all things in the environment. Rather, the world is "a structure or a unity of manifoldness. . . . The world is the structural whole which includes and transcends all environments."[12] Our sense of the world, correlated as it is with our sense of self, unites the moon's surface and the polluted Richmond air into a single reality within which the self finds itself and over against which the self defines itself. We *have* a world, while living *in* the world at the same time.

The self structures itself both within and against its world, and this gives the self its center: Tillich calls this a "perspective-center." Something perceives. And, at the same time, this something perceives that it is perceiving. Consciousness of the world redounds to elicit self-consciousness. There would be no self-consciousness without world-consciousness, and there would be no world-consciousness without self-consciousness as well. The two are relational correlates. Self and world come together in a single package.

The self is identifiable only by reference to the world, and vice versa. A self without a world is dead; a world without a self is empty. Hence, self and world are not simply two things that we can identify separately or analyze independently. Both self and world are real, to be sure; but they are real together.

The borders between the person and the external world are somewhat undefinable, shifting, and changing. What makes us persons is in large part our reliance upon the unifying self to maintain internal integrity in relation to an external world. However, what is being challenged by the puppet determinism of today's gene myth is this confidence in the internal integrity of the self. If we are determined by strings pulled by the DNA puppeteer, can we be reduced to biological determinants more internal to the person than even the self? Should we consider our DNA as part of our world—that is, as part of the environment within which the self finds itself? Should we think of DNA as internal to our body but not internal to our self? And, if our DNA internal to our body is ironically external to our self, and if DNA is determinative, have we lost our self entirely? Is the freedom belonging to our self, and hence to our soul, only a delusion?

I think not. We have one significant piece of observational data going for us. The sense of having an integral and unifying center in the self remains subjectively observable. This is fundamental to human experience, and virtually presupposed by all interpretations and explanations of human experience. To date no explanation based on the empirical study of genetic determinants has exhaustively reduced the experienced self to more primitive biological factors. The burden of proof remains on the shoulders of those who would have us believe that what we observe is in fact a delusion. Until such convincing proof is in, our internal observations of our selves engaging in the exercise of freedom can remain a fundamental datum for theologian and nontheologian alike.

Certainly the soul or self finds itself in a most precarious situation. It is utterly dependent upon its particular world for its very existence. Its particular world includes environmental determinants without and biological determinants within. Despite this precariousness, the human soul can dream dreams as yet undreamt and sing songs as yet unheard. It can draw its own world up into consciousness, and it can become conscious of this consciousness. Born on Earth its thoughts can soar toward heaven, and even beyond heaven to the stars, and beyond the stars to infinite possibilities. Theologically speaking, the human soul is being called constantly by God to go beyond its biological beginnings and beyond its cultural achievements to embrace new creation. The soul finds itself ever at the metaxy, at the point of tension between soil and spirit.

Freedom and Destiny

If our genome represents the world within, and if DNA is determinative, does this constitute necessity? And does genetic necessity diminish or dissolve freedom?

No. Freedom transcends necessity without destroying it. Freedom is not the enemy of necessity, nor of determinism for that matter. Ordinarily conversants in the freedom versus determinism debate speak about freedom against necessity, as if freedom were a form of indeterminism. This misformulates the issue. The opposite of necessity is possibility, not freedom. When freedom and necessity are set over against one another, the presumption is typically that necessity should be understood as mechanistic determinism and freedom understood as indeterministic contingency. The problem is that this presumption does not square with our actual experience of freedom. What we experience as human beings is the person as the bearer of freedom within a larger structure of the world to which the self belongs.

With this observation in hand, Tillich searches out an alternative formulation that better squares with human experience. He sets freedom over against, yet in complementary relationship to, destiny. Destiny points to the situation in which we find ourselves facing a world to which we also belong. Because destiny is a dynamic concept, it connotes more than simply the world understood spatially as that which is outside the self. Destiny is the inextricable otherness with which we interact over time as well as space, an interaction that centers us while we as centers of selfhood, act and react against it.

Quite frequently debates over human freedom invoke the concept of the will. Advocates of determinism and advocates of free will both typically assume that the human will is an object or a function or a thing, a thing that allegedly has freedom as a quality or attribute. But the will cannot be a thing. By definition, Tillich says, a thing is completely determined; therefore, the will could not have freedom if it is a thing. It is a mistake for indeterminists to presume that the will is an identifiable thing. The *freedom of a thing* is a contradiction in terms. If both debaters assume that the will is a thing, then the determinists deserve to win the debate.

What we experience as freedom is not a function of a thing such as will; rather, it is the function of a self, the expression of a human person who is not a thing but rather a self with a rational capacity to deliberate, make decisions, and shoulder responsibility. Tillich does not object to shorthand speech where we might use the term "will" to refer to the personal center, a form of synechdoche that represents the totality of the self. As long as we avoid the misleading presumption that the personal self is one thing and the will another, the phrase "free will" will be intelligible. It is better, of course, to speak of human freedom or the freedom of a person. This recognizes that every function of the person, including the cells of his or her body, participate in the constitution of the personal center, the human self or soul.

This implies further that human freedom cannot be located in something less than the self; it cannot be simply the result of a physical or biological process within our bodies. Human freedom is not just a version of physical indeterminism. No matter how much philosophical excitement was generated a half-century ago by the discovery of Heisenberg's Principle of Uncertainty at the subatomic level, many theologians insisted that we should not try to base our doctrine of freedom on unpredictable patterns of electron movement. Discovering indeterminacy in nature does not automatically mean that we can speak of human freedom.[13] It would seem that what is said here regarding the new physics would apply to the new genetics as well. Whether we discover indeterminacy in random mutations or even the opposite (genetic determinism as the gene myth pictures it), this would not automatically mean we could explain human freedom, or explain it away for that matter. Freedom as we human beings experience it lives at the level of the whole self; it is not reducible to the individual parts or processes that make up our physical bodies.

We are operating with the doctrine of holism here, according to which the whole is greater than the sum of the parts. A personal self is a whole exhibiting an integrity that is greater than the totality of its constituting parts, whether those parts be DNA or socializing experiences. Tillich puts it this way: "Ontologically the whole precedes the parts and gives them their character as parts of a special whole. It is possible to understand the determinacy of the isolated parts in the light of the freedom of the whole—namely, as a partial disintegration of the whole—but the converse is not possible."[14] In short, freedom is expressed and experienced by the whole person, not the person's individual parts.

Borrowing Tillich's nomenclature, we can say we experience freedom in the form of deliberation, decision, and responsibility. The first, *deliberation*, points to a mental process of weighing or evaluating arguments and motives. The personal self who does the weighing is independent of the weighing process itself. Deliberating is an activity of a human subject, and when weighing various motives, the subjective self presumes it is free from them. The centered self does the weighing and reacts as a whole to the struggle between motives.

The second, *decision*, reminds us of its sister word, "incision." This connotes the image of cutting. Among other things, a decision cuts off possibilities. The possibilities cut off must have been real possibilities, otherwise no cutting would have been necessary. The person who engages in the act of cutting must be beyond what he or she cuts off and excludes.

By deliberating and deciding we cut through "the mechanisms of stimulus and response."[15] Our personal center has possibilities, but we are not identical to them.

The third, *responsibility*, points to the obligation of a person who has freedom to respond if questioned about his or her decisions. The individual cannot ask anyone else to answer for him or her. Because actions that result from deliberation and decision are determined neither by something outside the person nor by any part within the person, the centered whole or self shoulders the responsibility for such actions.

These observations make the freedom-destiny polarity understandable. Tillich again: "Our destiny is that out of which our decisions arise; it is the indefinitely broad basis of our centered selfhood; it is the concreteness of our being which makes all our decisions *our* decisions. When I make a decision, it is the concrete totality of everything that constitutes my being which decides. . . . Destiny is not a strange power which determines what shall happen to me. It is myself as given, formed by nature, history, and myself. My destiny is the basis of my freedom; my freedom participates in shaping my destiny."[16] Freedom and destiny belong together. Destiny does not point to the opposite of freedom; it points rather to freedom's conditions and limits and products. If Tillich were alive today, he would probably place our genomes in the category of destiny. As destiny, our genetic inheritance would provide both a limit as well as a resource for deliberation, decision, and responsibility.

The mystery and drama of human freedom begins with the ability of the personal self to transcend the world to which it inextricably belongs and, further, to engage in self-transcendence. Our human freedom is a finite freedom. It has limits. Yet, through language and thought, our minds can imagine transcending those limits. The human imagination invents philosophy and projects universals that liberate us from bondage to the concrete situation to which our biological natures are subjected. We are free to build and play in imaginary structures above the actual structures to which we are by nature bound. We are free to create new worlds, to develop arts and technologies that so alter our environment, and even our biological make-up, that our world of meaning transcends the dictates of our prehuman natural history. Finally, we have the freedom to contradict ourselves, even to the extent of surrendering our freedom. We are free to separate ourselves from our essential nature, to dehumanize ourselves, to sell ourselves into slavery. We are free to enter into sin.

Destiny and Definition

The concept of the human being with which we are working here is not a static one. The definition is not fixed. Rather, *we are on the way:* we are becoming human. The history and future of nature participate in this defining process, a process that is not yet complete. Our destiny contributes to the future defining process.

God will have a hand in this future destiny. In addition to biological determinism and environmental determinism perhaps we should add divine determinism. These do not remove human freedom, to be sure; but they contribute to the definition of humanity as a whole and to the identity of each of us individually.

Decisive among these determinants is divine destiny, which brings anthropology within the purview of the Christian perspective. The two basic categories with which a specifically Christian anthropology works are the image of God (*imago dei*) and human sin. Especially relevant here is the concept of the *imago dei*, which we might interpret evolutionarily and eschatologically in terms of Christ as the New Adam. What this means is that we are *becoming* human. We are not born that way. This applies to the whole human race as well as to individuals. We will become what we truly are only in the fulfillment of history, only at the arrival and fulfillment of the promised kingdom of God.

Human identity looks to the future, not the past. Along with Wolfhart Pannenberg we may find we have to reject the naive orthodox view that in the past there once existed a Garden of Eden with perfect people in it. "The idea that there was an original union of humankind with God which was lost through a fall into sin is incompatible with our currently available scientific knowledge about the beginnings of the race."[17] In the event that DNA studies should actually show that the present human race does in fact derive from a single primal pair of parents, and if this were to look like science is supporting the Adam and Eve story, the Pannenberg position would not necessarily be undercut. The weight of this objection to a historical Eden falls on the idea of a past perfection, not a past per se. When we think of human definition in terms of the New Adam rather than the Old Adam, we look to the future for perfection, not the past.

The *imago dei*—the image of God in us—belongs to our destiny (*Bestimmung*), which is calling us forward toward God and in the process is actually engaged in our formation (*Bildung*). Humanity in history is in the process of transcending itself, of gaining its own identity, of attaining its true selfhood. Because selfhood is still in the process of coming to be, we cannot say that who we are is due strictly to self-initiation. We do not attain self-realization in a promethean manner on the basis of our own power alone. We have help. That help comes in mediated form from interaction with other people around us, from our language and culture, and finally from God. It comes ultimately from the divine *imago* which is drawing us toward fulfillment. Here we close the circle. The metaxy between soil and spirit constitutes our openness to God, while at the same time it is also God calling us.

This calls for a response of faith. Faith, as Luther described it, is trust in something *extra se in Christo* (beyond ourselves in Christ). We abandon ourselves and build our lives upon that in which we place our trust. Trust recognizes and exploits our dependence upon that to which we abandon ourselves. Pannenberg follows Cicero in saying that we are religious by nature, whether we acknowledge it or not, because we live on the basis of a fundamental trust which sustains our life.[18] Christian faith consists in consciously and actively putting our trust in the God who is *extra nos* yet who—through this very faith—is in the process of drawing us toward our true selves.

This makes our personhood and personality essentially future-oriented. "The person lives by the future in which its trust is placed."[19] But, we might ask, how do we know who we are in the present moment? Answer: the identity which we experience at the present moment—the moment within history and prior to its consummate end—is accomplished by our time-bridging consciousness. We remember the past and anticipate the future, all the while envisioning our own particular part in the drama of the wider world.

What makes such time-bridging consciousness possible is our participation in spirit. "Personality is to be understood as a special instance of the working of the spirit," says Pannenberg.[20] The spirit provides the continuity of our identity over time. Similarly, it is the spirit which provides continuity between all things temporal and spatial. It is the power which is in the process of integrating the parts of God's creation into a single comprehensive whole, the new creation. Because it belongs to the final future, we think of it as eschatological. The whole will be complete in the eschatological future.

Resurrection and the Spiritual Body

The future whole symbolized by the new creation includes resurrection. Our future resurrection is our destiny, and it contributes to the definition of our humanity. The biblical description of resurrection recalls the soil-spirit tension when it prophecies our rising as a spiritual body.

The New Testament foresees both discontinuity and fulfillment. Discontinuity comes with death. Death is total. Nothing survives. St. Paul uses the analogy of the seed, which looked to him as if it were dead. "What you sow does not come to life unless it dies," he writes (1 Corinthians 15:36). It would follow from this that nothing in the DNA is the trigger for immortality. Genetic determinism dies with us.

Continuity in the form of fulfillment comes with transformation. Paul, still using the seed analogy, tells us that what is sown perishable is raised imperishable. What is sown in dishonor is raised in glory. What is sown in weakness is raised in power. Then he adds that what dies as a physical body is raised a spiritual body (1 Corinthians 15:42–44). The Greek term for physical body is actually *soma psychikon*—that is, an ensouled body. The term for the resurrected body is fascinating, *soma pneumatikon*—literally, a spiritual body. This is a curious term to someone who habitually distinguishes body from spirit. So, what could Paul mean? Paul cannot mean a spirit without a body, because it is a spiritual *body*. Yet, Paul distinguishes it from the ensouled body, which I take to be referring to the soil and spirit existence we currently enjoy. Perhaps Paul means that body and spirit are so melded together that the metaxy is removed, that the tension is resolved, that a new state of harmony is achieved. Perhaps the spiritual body of the resurrection is the fulfillment of the human soul currently making its way through the struggles of soil and spirit.

Normally we do not ask molecular biologists to search the DNA looking for the trigger to immortality. Even if genes for longevity are eventually located and pharmaceuticals lengthen human life to two or three centuries, we still would not have what the Bible means by resurrection. Our resurrection from the dead will be a divine act, an action of God's grace that accompanies the advent of the new creation. The Christian concept of resurrection is not intended to provide an escape from death, or even an opportunity for the individual soul to slip off into a heaven of souls. Resurrection belongs to God's promise of redemption, of transformation of the present creation into the new creation. That the advent of the new will be due to a divine action and not the continuation of evolutionary progress marks the discontinuity. That the advent of the new creation will bring transformation and wholeness marks the future fulfillment of nature and history as we know it. This is our future destiny, something the theologian includes in the definition of our true humanity.

Divine Freedom and Human Freedom

Now to the question: In light of future freedom and future destiny, how can we think of God's contribution to human freedom? My answer: human freedom is derived from divine freedom. The neo-orthodox giant Karl Barth, answers: "God in His own freedom bestows human freedom.... Human freedom is the gift of God in the free outpouring of His grace."[21]

When we opened this chapter we noted how many theologians begin with the assumption of a fixed pie of power in the universe. If God's slice of the power pie gets too large, then the human slice gets too small. And if God the omnipotent one takes the whole pie, then we humans have no power left to determine our own future. Omnipotence eliminates human freedom, according to this scheme. Our human freedom seems to be denied just as a serf's freedom is denied by the power of his or her lord. This oppressiveness could lead to a promethean call for liberation, a revolt against God.

The idea that God oppresses us with divine power is totally alien to the Christian theologian. God loves, not oppresses. So there must be a solution to this problem. The fixed-pie theorists in the history of Christian thought have on occasion sought to preserve human freedom by introducing the concept of divine self-limitation. According to this principle, God the omnipotent one voluntarily withdraws divine claim to all the power. This opens up a power vacuum, so to speak, that is then filled by human energy and creativity. When God vacates, we enter. Where God is absent we enter with human freedom.

This is why Langdon Gilkey advocates "an explicitly ontological doctrine of the self-limitation in every present of the divine power in relation to the freedom of the creature."[22] John Polkinghorne adds that God's motive for self-limitation is love, wherein love seeks freedom and self-control in the creature. "This curtailment of divine power is, of course, through self-limitation on his part and not through any intrinsic resistance in the creature. It arises from the logic of love, which requires the freedom of the beloved. . . . God remains omnipotent in the sense that he can do whatever he wills, but it is not in accordance with his will and nature to insist on total control."[23] On this point, Wolfhart Pannenberg is in agreement. He asserts that God is the all-determining reality and, further, "this all-determining power is itself determined only by itself and not subject to determination by anything else, unless it determines that it should be determined by something else."[24]

Yet, this notion of divine self-limitation partly misses the point. It is built on the false assumption that divine and human power are in competition. The better assumption, it seems to me, is to see God's power and our human power operating on different but complementary levels. Divine and human power do not compete. Rather, divine power is the source of human power. God, by asserting the power to create a future, determines that we will be free. Instead of a fixed pie of power, God and the creation are engaged in an ongoing interaction that results in creative empowerment.

God is free. God gives freedom to us. God liberates. Human freedom in its fullest and finest sense is a free gift from a free God that we open, unpack, and enjoy.

Just how God delivers this gift of freedom to us is subtle. It comes to us with transformatory power. It is not built into our creaturely natures. We do not come off the assembly line with freedom as a factory feature. It is more like a dealer accessory yet, when added, it redefines the person we are driving. The human achievement and enjoyment of freedom belongs to God's work of redemption.

Human freedom understood as human self-movement entails the quality of personhood, of selfhood or soul. In the life of the trinitarian God we find aboriginal "personalness," and this personalness is contagious. It is creative of more personhood. Karl Barth says that God "is the one, original and authentic person through whose creative power and will alone all other persons are and are sustained."[25] Elsewhere he says, "As a gift of God, human freedom cannot contradict divine freedom."[26]

Key to apprehending God's freedom is appreciating the fact that God is in relationship, in relationship to us and to the created world. Barth makes this point: "God has the freedom to be present with that which is not God."[27] God is not stuck in absoluteness or unconditionedness. God can and does choose to involve the divine self in loving relationship.

Essential here is the fact that God's absoluteness does not require that God remain absolute in the sense of being unconditioned, untouchable, aloof. God is so free that God is not constrained even by the qualities of divinity. God is not constrained by what philosophical theologians list as a "bunch of omni's": omnipotent, omniscient, omnipresent, etc. God is so free that he can divest the divine self of omnipotence, omniscience, and omnipresence in order to enter into relationship with a concrete, conditioned, and finite world. God is so free that God and God alone defines divinity, and God does so by choosing an incarnate relationship with the created world. "If, then, the freedom of God is understood primarily as His own positive freedom," writes Barth, "it can and must be understood secondarily in His relationship to that which is other than Himself."[28]

This brings us to what we earlier identified as Christian freedom. The neo-orthodox theologians of Karl Barth's era have stressed again and again that the freedom about which we speak here is not merely freedom *from* but freedom *for*—that is, authentic freedom empowers and enables personal selves to reach out in action and achievement on behalf of the welfare of others. "Human freedom is only secondarily freedom *from* limitations and threats," writes Barth. "Primarily it is freedom *for*." This freedom for, the freedom which we as human beings receive and act out of, echoes or extends the freedom enjoyed within God's trinitarian life. "God is primarily free *for*; the Father is free for the Son, the Son for the Father in the unity of the Spirit."[29]

Freedom for, as we experience it, begins with liberation. And, liberation, understood as the bestowing of freedom by God, begins with redemption. It is liberation from sin.[30] What has taken place in the person of faith, in whom the Holy Spirit places the presence of the living Christ, is the inbreaking of a new depth of personhood making us at peace with God, at peace with our neighbor, and at peace with our inner self. The result is a life of joy.

Freedom for puts us on the doorstep of ethics, evangelical ethics. Walking through the door takes us into beneficence. Even though bioethicists define beneficence as serving the welfare of others when there is no risk of danger to the one bestowing such service, evangelical ethics does not mandate avoiding the risk. God took a risk in the incarnation. Those sensing that they have been liberated by this divine act are free to risk as well. This is a key element in Christian freedom.

Advent Shock and Proleptic Ethics

Eschatological wholeness stands in continuity with the present world because it is the fulfillment of nature and history. As fulfillment, however, it is also transformative. The symbol of the new creation implies transformation, to be sure; yet it is transformation of this world and our lives on behalf of our fullest self-realization.

Eschatological wholeness stands in discontinuity as well. The contrast between our present reality and God's future reality is sufficiently sharp to lead Yale theologian Letty Russell to coin the phrase, "advent shock." She begins by distinguishing the two Latin terms, *futurum* and *adventus*. According to *futurum*, the future is a continuation of conditions and directions of the present reality. It is an "extrapolation out of the continued groaning of a world in pain."[31] The pain we feel today from injustice (or disease) can be extrapolated for the indefinite future. *Adventus*, in contrast, begins with a vision of a radically different future brought about by God. It is a future that God has promised. The future of God will include a divinely established justice and the elimination of pain. It will be a new creation. The radicalness of this transformation creates a sense of shock. Living today out of a vision of God's future creates a sense of maladjustment to the present.

This maladjustment leads to a proleptic form of ethics—that is, taking creative and transformative action in the present stimulated by our vision of the future. "This maladjustment with the present is called *advent shock*. Because of advent shock we seek to anticipate the future in what we do, opening ourselves to the working of God's Spirit and expecting the impossible."[32] When we think of God's action in terms of newness, it is easy to think of human action as similarly aimed at the new.

When we put this vision of newness together with Christian freedom and its accompanying ethic of neighbor love, we get proleptic ethics.[33] To think proleptically is to think creatively. This is a form of ethical thinking and moral action that does not seek primarily to conform itself to a set of laws or commandments, although there can be some value in this. Nor does it pursue a higher life of virtue, wherein we cultivate within our selves greater sensitivity and appreciation for values such as beauty, truth, and justice (although there is great value in this too). Rather, proleptic ethics tends toward seeking the most practical way to love the neighbor in light of a vision of a better future. Because it is oriented toward the new, proleptic practice does not attribute sanctity to previously established norms and policies for getting the job done.

Applied to the Human Genome Project, proleptic ethics would seek to guide the research in the direction of beneficence, toward medical technologies that would contribute to the long-term health and flourishing of the human race (as well as the other organisms on our planet who share our genes). Genetic research should lead to new technology. New technology should lead to new therapy. New therapy should lead to better health. Science should serve the relief of human suffering.

When the gene myth says, "Thou shalt not play God," the proleptic ethicist will take this as a word of caution but not revere it as a "No Trespassing" sign. Warning buzzers against the promethean attempt to dominate nature for the sheer purpose of gaining technological control need to be sounded daily as the research proceeds. Warnings should definitely be heeded in the case of germline intervention and enhancement.

Cautions and warnings against playing God should be heeded, but they should not be allowed to intimidate. Nature, as complex and wondrous as it is, is not in itself sacred. To be sure, nature is far more than just a tool to be used, far more than merely an object of instrumental value. Nature is in us. We are natural, fully natural. We need to respect and honor nature as we respect and honor ourselves. But we need also to notice that nature is not static. It moves. It changes. It evolves. Neither its past nor its present constitute an unchangeable sacred. We do nature no service by hallowing its past or present state. As soil and spirit in a world of dynamic nature, the direction of the created cocreator is forward.

Another set of warning strobes should startle our eyes. As we focus on those in the future who will benefit from the advances of genetic science and medical technology, we should not exclude the others from our horizon of vision. The near utopian promise of genome research is the future ability to guide the genotypes of many individuals toward greater health and wellbeing. The cries of near perfect children—conforming to a culturally conditioned image of the perfect child—will be heard in maternity wards. Their prospects for living their entire life without suffering from cancer or Alzheimers or cystic fibrosis will be measurably increased. Medical science will have delivered a new standard of health care and raised our norm for acceptable human well-being. This will be the focus.

At the frame and easily missing from our visionary picture will be those who do not benefit. Due perhaps to their suffering from a disease yet without a cure, or due perhaps to their socioeconomic class that prevents them access to the new health care, many among us will wander our streets with disabilities. Children born the old fashioned way, without benefit of genetic engineering or selection, will live among us for the foreseeable future. Their genomes will receive what the roll of the DNA dice yields. Some will approximate the perfect child. Most will not. It would be a grave injustice indeed if the shadow side of genetic science and medical technology created new forms of neglect, injustice, and marginalization. Proleptic ethics keeps our eyes directed toward the shadows as well as the brightly lit possibilities.

Conclusion

The task of this book has *not* been to rally the crowd one more time to give a cheer for freedom. Rather, we have tried to give answer to the modest question: Does the new knowledge deriving from the frontier of genetic research confront us with a genetic determinism that threatens our confidence in human freedom? We have seen that the answer has been no. The question itself is based upon a category mistake. The concept of freedom applies to the whole of a person, not to one of his or her parts, even a genetic part. Freedom is exercised at the level of the person, the self, and it takes the form of deliberation, decision, and responsible action. Determinism at the genetic level does not obviate free will at the person level. Genetic determinism just like all conditions of finitude place each person in his or her particular situation, readying the person to exercise freedom.

Yet we must address the question because it is being posed consciously or unconsciously as we interpret scientific findings through the conceptual set we have identified as the *gene myth*. Of the four freedoms we have reviewed, political liberty seems to be the primary operative model in the gene myth, although natural freedom and future freedom are also present. Just as an oppressive government takes away our political freedom, so also it appears that genetic expression within each of our bodies may be doing the same. The model of external coercion that denies our circumstantial freedom is applied analogically by the gene myth to something internal. Our DNA starts to look like a puppeteer pulling the strings that make us dance. Instead of the puppet deliberating and deciding and acting, we can imagine the DNA taking over these responsibilities. We can imagine that the freedom we think we have is in reality a delusion, a sham perpetrated on our consciousness by a selfish gene that manipulates us for its own purposes.

Ignored by the gene myth is the concept of moral freedom, wherein the self orients itself to a good that transcends the self. Rather, it presupposes that the internal loss of political freedom accounts for the loss of free will at the personal level. Although the loss of political liberty is the model whereby the loss of free will is interpreted, the gene myth also works with a variant of future freedom.

Future freedom is affirmed by the gene myth's implicit prometheanism. The determiner of the future of the human race will be the human race, and we humans will use scientific knowledge and technological innovation to engineer ourselves for that future. As Prometheus stole the secret of fire from the gods, we will steal the secrets of life from DNA. Once we have the knowledge, we will have the power. And with this power we can do what? Can we do damage? Yes, we can do damage beyond measure. Without any commitment to moral freedom, future freedom quickly becomes fearsome.

Fearing a future without moral commitment, the gene myth produces antiprometheans right along with prometheans. The fear of the antiprometheans leads to the posting of "no trespassing" signs, to proscriptions against technologically profaning an implied sacredness to nature. Some signs read: "Don't Play God!"

At hearing the word "God" in this desperate cry for protection of nature, the ears of at least a few theologians are perking up. After trying to discern whatever theological content the phrase playing God might have, it appears that it has very little. Its primary function is to serve as a protective shield against a technological threat to an assumed sacredness of DNA or of nature in general. Having relatively little cognitive content, "Thou shalt not play God" comes from the voice of panic amidst the overwhelming flood of new knowledge and new opportunities and new vested interest groups and new potentials for injustice. As a warning against foolish prometheanism, we should heed it.

This prompts a serious theological and ethical question: How should we play human? How should we understand our human relationship to God and to nature and to future generations of God's creatures? I recommend that we take advantage of the ability to deliberate, decide, and take responsible action indicative of the freedom that the gene myth did not in fact take away from us. I recommend that we orient our free wills around the good, the long-range good for the human race and for life on our planet as a whole. I recommend beneficence.

I offer these recommendations for theological reasons. The first reason is anthropological. At this point, the most helpful way to think of the image of God in us, the *imago dei*, is to think of the human race as God's created cocreator. We have been created by God. Yet, God created us to be creative. The creation is ongoing. Nature is dynamic. We are growing. Everything is in process. We cannot stand still. To be responsible we should employ science and technology and politics to exert whatever direction we can to guide nature's movement as well as human history toward the well-being and flourishing of all God's creatures.

The second reason is eschatological and proleptic. We can catch a glimpse of the good as we catch a glimpse of God's promised future, the kingdom of justice. When John of Patmos saw a vision of the New Jerusalem descending, he saw a new creation wherein God will find a home among mortals, wherein God "will wipe every tear from their eyes. Death will be no more; mourning and crying and pain will be no more, for the first things have passed away" (Revelation 21:4). This is God's promise. It may be difficult to believe, yet this is the promise. On the basis of the promise we can paint visionary pictures of a transformed world. We can set our sights toward a divine future. The value of this vision is that it provides an image of the good accompanied by the hope that energizes work to embody part of that vision ahead of time. Although many ambiguities and dangers remain as we explore genetic science, one thing must be said: genetic medicine promises a significant measure of potential for relieving crying and pain and mourning for numerous individuals. When pressed by moral freedom into the service of the good, genetic research and medical technology have proleptic value.

Research science has its own intrinsic value, to be sure. Probing the mysteries of the natural realm and becoming privy to her magnificent secrets is in itself a worthwhile vocation, needing no additional moral confirmation. But when placed within culture and society where the needs of the neighbor for better health and greater well-being become obvious, then science should serve technology which in turn should answer God's call to be creative and transformative, to make life qualitatively better for God's creatures.

< APPENDIX A >

Statement on the Gay Gene Discovery

Meeting in Berkeley, California, on August 17, 1993, the geneticists, ethicists, and theologians working together on a research project conducted by the Center for Theology and the Natural Sciences at the Graduate Theological Union dealing with "Theological and Ethical Questions Raised by the Human Genome Initiative" framed the following statement:

The recent report of an X-linked predisposition to some forms of male homosexuality has already created a furor, and we think it appropriate to pause and reflect critically if briefly upon what implications may or may not be justified. Although many critics have emphasized the high quality of the researchers and of the design of the research, they also point out that the results are still preliminary. Assuming that the research is replicated and supported by additional experimental designs, the scientific fact still does not itself determine the direction of the ethical interpretation of that fact. The ethical logic could go in several directions. On the one hand, genetic predisposition may remove sexual preference from the list of behaviors that can be blamed on individual choice or on the behavior of parents. On the other hand, it may lead to a cry to develop "cures" or eugenic measures to eliminate the predispositional gene, as have earlier studies of diversity between races lent misplaced credibility to discriminatory practices.

The possible genetic predisposition to some forms of homosexuality is but one instance of a broader theme of naturally occurring human genetic diversity. The problem of how to recast notions of freedom and responsibility in light of that diversity remains a basic philosophical issue. The honest expression of diverse opinions within the lesbian and gay community to the recently publicized claims shows the kind of maturity that we appreciate and commend to other social commentators. As theological and scientific commentators, we urge that the best of genetic research be taken seriously. There is a rich variety of theological reflection both in the religious traditions and in recent developments which leads us to affirm that the theological questions raised by this genetic research are profound and subtle. We recognize that genetic diversity requires a response of love, respect and justice.

IAN BARBOUR, Professor Emeritus of Religion and Science, Carleton College

PAUL W. BILLINGS, Chief, General Internal Medicine, Palo Alto Veterans Memorial Medical Center, and Associate Clinical Professor of Medicine, Stanford University Medical School

DEBORAH BLAKE, Professor of Religious Studies, Regis University, Denver

R. DAVID COLE, Professor Emeritus of Molecular and Cell Biology, University of California at Berkeley

RONALD COLE-TURNER, Professor of Theology, Pittsburgh Theological Seminary

LINDON EAVES, Distinguished Professor of Human Genetics and Professor of Psychiatry, Medical College of Virginia

PHILIP HEFNER, Professor of Systematic Theology and Director of the Chicago Center for Religion and Science, Lutheran School of Theology at Chicago

SUZANNE HOLLAND, graduate student, Graduate Theological Union, Berkeley

SOLOMON KATZ, Professor of Biological Anthropology, University of Pennsylvania, Philadelphia

KAREN LEBACQZ, Professor of Christian Ethics, Pacific School of Religion and Graduate Theological Union, Berkeley

DONNA MCKENZIE, graduate student, Graduate Theological Union, Berkeley

ARTHUR PEACOCKE, Warden, Society of Ordained Scientists, Exeter College, Oxford, England

TED PETERS, Professor of Systematic Theology, Pacific Lutheran Theological Seminary and Graduate Theological Union, Berkeley

ROBERT JOHN RUSSELL, Professor of Theology and Science and Director of the Center for Theology and the Natural Sciences, Graduate Theological Union, Berkeley

THOMAS A. SHANNON, Professor of Religion and Social Ethics, Worcester Polytechnic Institute, Worcester, Massachusetts

< APPENDIX B >

Playing God
with David Heyd

Philosopher David Heyd at the Hebrew University of Jerusalem embraces a variant of the playing God ethic. He advocates that we play God. He begins with many of the concerns taken up in this book, but what he advocates is more promethean.

Heyd is aware of how the phrase "playing God" gets invoked when we scold our scientists or excoriate our society for tampering with life-giving natural processes. To accuse someone of playing God is to condemn them for unauthorized human intervention into a process within nature that has been divinely ordained. It appears to be an assertion of human pride, an arrogant transgression. "However," argues Heyd in a most fascinating way, "if indeed the capacity to invest the world with value is God's image, it elevates human beings to a unique (godly) status, which is not shared by any other creature in the world. This is playing God in a creative human-specific way."[1] Obviously Heyd is alluding to Genesis 1:26–31, wherein the human race is created in the divine image. If we human beings bear God's image, and if God is creative, then we are being godly when we are creative. Then Heyd adds a secular or nonreligious premise: metaphysically speaking, God does not exist. We live in a Godless world. Once we have denied existence to God, then we human beings are left with the task of playing God's role in the world.

Is Heyd then advocating a reckless prometheanism, a paving over of the planet with technological intervention that snuffs out all sprouting of natural processes? Promethean perhaps, but not reckless. His agenda is much more that of the ethical theorist asking a modest question: How can we ground value? We human beings are the creators of value. In a Godless world with no transcendent design or purpose, we human beings are the value begetters without which the world would remain valueless.

Heyd is tackling the question that futurists and intergenerational ethicists frequently ask: Just what kind of moral status can we attribute to future people who do not yet exist? To grant people who do not yet exist moral status, we would have to work within the framework of an impersonal ethic. If the persons in question are not actual, it follows that we would have to appeal to something impersonal such as duty or sense of the good or utility or whatever. But Heyd is unhappy with an impersonal ethic. He wants ethics to be grounded in persons, actual persons. His is a "person-affecting" approach. So, when it comes to the morality of making decisions today that will or will not bring new people into existence tomorrow, he focuses on the persons making the decisions. His ethic is generocentric—that is, it focuses on the generators or creators. Or, to use common parlance, it focuses on the parents rather than the children yet to be conceived. Hence it is parentocentric too. He dubs his work "genethics" because it applies to the genesis of future people. He defines "*genethics* as the field concerned with the morality of creating people, that is, decisions regarding their existence, number, and identity."[2]

His point is that value is created by human agents who already exist, not by potential people who do not yet exist. The decision to create—actually the decision to procreate—new people is analogous to God's creating the cosmos *ex nihilo*, out of nothing. We create valuers. Through procreating future persons we bring into existence new human subjects who will value an otherwise valueless world. The transformation of God's power of creation into the human power of procreation is the means of spreading the image of God in the world. "In a Godless world, in which procreation cannot be considered a religious injunction to take part in a divine scheme, we ourselves are gods, constituting value in the world by extending ourselves and our image through the mediation of our descendants. The logic of creation and procreation is the same; it is only on the metaphysical level, in which we reach the boundaries of value, where the distinction between the religious and the secular views become relevant."[3]

That the distinction between the religious and secular might become irrelevant is a prospect I applaud. That we should begin now to ponder our responsibility for future generations via genethics I also applaud. With regard to who plays God for the future, however, I differ. I believe that God plays God, and that we humans should play human.

Notes

Preface

1. Langdon Gilkey, *Naming the Whirlwind: The Renewal of God-Language* (Indianapolis and New York: Bobbs-Merrill, 1969) 234.
2. Carl Sagan, *The Demon-Haunted World* (New York: Random House, 1995) 35.
3. Sagan, *Demon-Haunted World*, 34.
4. Anne M. Clifford, "Creation," *Systematic Theology: Roman Catholic Perspectives*, ed. Francis Schüssler Fiorenza and John Galvin, 2 Volumes (Minneapolis: Fortress Press, 1991) I:239.

Chapter 1
Playing God with DNA

1. "The end product of the project is: a reference map (and sequence)—a source book for human biology and medicine for centuries to come." Victor A. McKusick, "The Human Genome Project: Plans, Status, and Applications in Biology and Medicine," *Gene Mapping*, ed. by George J. Annas and Sherman Elias (New York and Oxford: Oxford University Press, 1992) 18.
2. "Some 5,000 heritable disorders have been clinically characterized." C. Thomas Caskey, "DNA-Based Medicine: Prevention and Therapy," *Code of Codes*, ed. Daniel J. Kevles and Leroy Hood (Cambridge: Harvard University Press, 1992) 114–115.
3. Leroy Hood, "Biology and Medicine in the Twenty-First Century," *Code of Codes*, 158.
4. The most widely used method of gene transfer is to put the new gene aboard a retrovirus and send it into the cell. The problem is that retroviral integration procedures show little target-site specificity, and haphazard insertions into a patient's chromosomes can cause problems such as shutting down the wrong genes or even activating cancer genes. This is no reason for a scientific faint heart, because hurdles are being jumped. See: Frederic Bushman, "Targeting Retroviral Integration," *Science*, 267 (March 10, 1995) 1443–1444; and Jacqueline Kirchner, Charles M. Connolly, and Suzanne B. Sandmeyer, "Requirement of RNA Polymerase III Transcription Factors for in Vitro Position-Specific Integration of a Retroviruslike Element," *Science*, 267 (March 10, 1995) 1488–1491.

5. John Maddox, "Has Nature Overwhelmed Nurture?" *Nature*, 366 (November 11, 1993) 107.
6. Against genetic fatalism, Joseph S. Alper and Jonathan Beckwith argue that it is a mistake to believe (1) that genetically controlled or highly heritable traits are unchangeable; (2) that genetic traits should not be changed on grounds that Darwinian natural selection has sanctified them; and (3) that genetically based traits are more difficult to change than environmentally based ones. "Genetic Fatalism and Social Policy: The Implications of Behavior Genetics Research," *Yale Journal of Biology and Medicine*, 66 (1993) 511–524.
7. Dorothy Nelkin and M. Susan Lindee, *The DNA Mystique: The Gene as a Cultural Icon* (New York: W.H. Freeman and Company, 1995).
8. Ruth Hubbard and Elijah Wald, *Exploding the Gene Myth* (Boston: Beacon Press, 1993).
9. Walter Gilbert, "Towards a Paradigm Shift in Biology." *Nature*, 349 (1991) 99.
10. L. Jaroff, "The Gene Hunt," *Time*, 133:11 (March 20, 1989) 67.
11. Ted Peters, *Futures—Human and Divine* (Louisville: Westminster/John Knox, 1977).
12. President's Commission for the Study of Ethical Problems in Medicine and Biomedical and Behavioral Research, Morris B. Abram, Chairman, 2000 K Street NW, Suite 555, Washington, DC 20006, *Splicing Life* (November 1982) 54.
13. Thomas A. Shannon, "Cloning, Uniqueness, and Individuality," *Louvian Studies*, 19 (1994) 283–306, p. 299.
14. When a surgeon picks up a scalpel and tries to save a life or when a surgeon decides to withdraw a life support system, it is a form of playing God. Leroy Augenstein, *Come, Let Us Play God* (New York: Harper and Row, 1969) 12. Whether positively or negatively construed, the simple power of life and death seems to put us on God's doorstep.
15. Ernlé W.D. Young, *Alpha and Omega: Ethics at the Frontiers of Life and Death* (New York: Addison-Wesley, 1989) 1–6.
16. Philip Elmer-DeWitt, "The Genetic Revolution," *Time*, 143:3 (January 17, 1994) 46–53.
17. *Splicing Life*, 95–96.
18. Pope John Paul II, Discourse to those taking part in the *81st Congress of the Italian Society of Internal Medicine and the 82nd Congress of the Italian Society of General Surgery*, October 27, 1980: *AAS* 72 (1980) 1126.
19. See: Jane Goodfield, *Playing God: Genetic Engineering and the Manipulation of Life* (New York: Random House, 1977) 6.
20. Jeremy Rifkin, "Playing God with the Genetic Code," *Threshold*, 6:3 (January 1994) 17–18. Obtained from Student Environmental Action Coalition, P.O. Box 1168, Chapel Hill, NC 27514–1168.
21. Michael Crichton, *Jurassic Park* (New York: Ballantine, 1990).
22. *Human Life and the New Genetics*, A Report of a Task Force commissioned by the National Council of the Churches of Christ in the U.S.A., 1980, 41.
23. United Methodist Church Genetic Science Task Force Report to the 1992 General Conference, 114.
24. Burke Zimmerman, "Human Germ-Line Therapy," *Journal of Medicine and Philosophy*, 16:6 (1991) 606.
25. Robert L. Sinsheimer, "Genetic Engineering: Life as a Plaything," *Technology Review*, 86:14 (1983) 14–70.
26. Jeremy Rifkin, *Algeny* (New York: Viking, 1983) 17.
27. Rifkin, *Algeny*, 252. Although the phrase "play God" has been with us for some decades as a reference to the prospect of scientific creation or manipulation of

life, Jeremy Rifkin thrust it before the public with his 1977 book titled *Who Should Play God?* (New York: Dell, 1977).

28. Walter Truett Anderson, *To Govern Evolution* (New York: Harcourt, Brace and Jovanovich, 1987) 9, 135.

29. Ronald Cole-Turner, *The New Genesis: Theology and the Genetic Revolution* (Louisville: Westminster/John Knox, 1993) 45.

30. Cole-Turner, *The New Genesis*, 45.

31. Hessell Bouma, et.al., *Christian Faith, Health, and Medical Practice* (Grand Rapids, Michigan: William B. Eerdmans, 1989) 4–5. James M. Gustafson makes "the theological point that whatever we value and ought to value about life is at least relative to the respect owed to the creator, sustainer, and orderer of life." James M. Gustafson, "Genetic Therapy: Ethical and Religious Reflections," *Journal of Contemporary Health Law and Policy*, 8 (1992) 196. For Gustafson, the central question around which the issue of germline intervention is oriented is this: How do we define what is naturally normal for human life? For the theologian to answer this question, more than knowledge of biology is required. Required also is awareness of the divine ordering of human life.

32. Ted Peters, *GOD—The World's Future* (Minneapolis: Fortress, 1992) chapter 4.

33. Philip Hefner, "The Evolution of the Created Co-Creator" in *Cosmos as Creation: Science and Theology in Consonance*, ed. by Ted Peters (Nashville: Abingdon, 1989) 212; and Hefner, *The Human Factor: Evolution, Culture, and Religion* (Minneapolis: Fortress, 1993) 35–42. See also James M. Gustafson, "Where Theologians and Geneticists Meet," *Dialog*, 33:1 (Winter 1994) 10.

34. Mortimer J. Adler makes this point when delineating the first three of these types of freedom: natural freedom, moral freedom, and circumstantial freedom. *Six Great Ideas* (New York: Macmillan, 1981) 140–142. See his *The Idea of Freedom: A Dialectical Examination of the Conceptions of Freedom*, 2 vols. (Garden City, NY: 1958–1961).

35. Charles Taylor, *Sources of the Self: The Making of Modern Identity* (Cambridge: Harvard University Press, 1989) 3.

36. Taylor, *Sources of the Self*, 34.

37. "We say that we have a free choice because we can take one thing while refusing another ... the proper object of choice is the means to the end, and this, as such, is in the nature of that good which is called useful." Thomas Aquinas, *Summa Theologica*, I:Q.83:Art.3. Like Aristotle before him, Thomas sees choice as motivated by reason (beliefs) and appetites (desires).

38. Thomas Aquinas, *Summa Theologica*, II:ii:Q.183:A.4.

39. Augustine, *The Spirit and the Letter*, 5.

40. Martin Luther, "The Freedom of a Christian," in *Luther's Works*, Vols. 1–30, edited by Jaroslav Pelikan (St. Louis: Concordia, 1955–1967) and Vols. 31–55, edited by Helmut T. Lehman (Minneapolis: Fortress, 1955–1986) 31:344.

41. Luther, *Luther's Works*, 31:351.

42. Luther, *Luther's Works*, 31:365.

43. Letty M. Russell, *Human Liberation in a Feminist Perspective—A Theology* (Louisville: Westminster/John Knox, 1974) 20.

44. Russell, *Human Liberation*, 30, italics in original.

45. Thomas A. Shannon makes it clear that beneficence understood as a duty to help others is required of us only "when we can do this without risk to ourselves." Thomas A. Shannon, ed. *Bioethics*, (New York: Paulist, 3rd ed., 1987) 7–8. Christian freedom expressed through beneficence would support such a duty, to be sure, but proceed beyond by eliminating the protection against risk to ourselves.

46. Tom L. Beauchamp and LeRoy Walters, *Contemporary Issues in Bioethics* (Belmont, CA: Wadsworth, 4th ed., 1994) 25.
47. Ronald Munson, *Intervention and Reflection: Basic Issues in Medical Ethics* (Belmont CA: Wadsworth, 1992) 34.
48. James F. Childress and John Macquarrie, editors, *The Westminster Dictionary of Christian Ethics* (Louisville: Westminster/John Knox, 2nd ed., 1986) 57.

Chapter 2
Puppet Determinism and Promethean Determinism

1. Richard C. Lewontin, Steven Rose, and Leon J. Kamin, *Not In Our Genes: Biology, Ideology, and Human Nature.* (New York: Pantheon, 1984) 6.
2. Lewontin, *Not In Our Genes*, 75. "every organism is the unique product of the interaction between genes and environment at every stage." Ibid., 95.
3. Lewontin, *Not In Our Genes*, 73.
4. *USA Today* (June 28, 1995) 1.
5. R. David Cole, "Genetic Predestination?" *Dialog*, 33:1 (Winter 1994) 20–21.
6. Cole, "Genetic Predestination?" 19.
7. Maddox, "Has Nature Overwhelmed Nurture?" 107.
8. Cole, "Genetic Predestination?" 21.
9. Gilbert, "A Vision of the Grail," *Code of Codes*, 96.
10. Francisco J. Ayala, "The Difference of Being Human," in *Biology, Ethics, and the Origins of Life*, ed. by Holmes Rolston III (Boston and London: Jones and Bartlett Publishers, 1995) 122.
11. Alfred North Whitehead, *Science and the Modern World* (Cambridge: Cambridge University Press, 1926, 1953) 64, and *Process and Reality* (New York: Macmillan, Free Press, 1929, 1978) 7–8.
12. Behavioral genetics along with evolutionary psychology represent nonmolecular approaches to genes. "The distinction between molecular and nonmolecular studies is important," writes Joseph S. Alper and Jonathan Beckwith, because "the methodologies are quite different. Since the molecular studies involve a search for the genes themselves, certain aspects of the technical analysis and interpretation are relatively straightforward. In contrast, nonmolecular studies rely on less direct evidence for the existence of genes correlated with behavior." "Genetic Fatalism and Social Policy: The Implications of Behavior Genetics Research," *Yale Journal of Biology and Medicine*, 66 (1993) 513.
13. Heritability does not describe an individual person who is 62 inches tall as having 50 inches due to genes and 12 inches due to environment. Rather, heritability measures the variance of a trait present in a population due to genetic differences divided by the total variance in the trait. In some instances "heritable trait" is synonomous with "genetic trait," but a pronounced genetic trait in an individual may not have high heritability in the population.
14. T.J. Bouchard, Jr., D.T. Lykken, M. McGue, N.L. Segal, and A. Tellegen, "Sources of Human Psychological Differences: the Minnesota Study of Twins Reared Apart," *Science*, 250 (1990) 223–228.
15. John Horgan, "Eugenics Revisited: Trends in Behavioral Genetics," *Scientific American*, 268:6 (June 1993) 122–131.
16. Lewontin, Rose, and Kamin dispute the methods and conclusions of behavioral genetics. They claim that investigators in such studies depend too heavily on the verbal accounts of volunteer twins to provide details about the conditions and duration of their separation. Evidence exists that twins sometimes tend romantically to exaggerate the degree of their separation, and alleged "facts" reported by the twins have sometimes been mutually contradictory. These problems are

overhwelming, and Lewontin and colleagues try to account for traits with greater appeal to environment. *Not In Our Genes*, 93–119.

17. Lawrence Wright, "Double Mystery," *The New Yorker*, 71:23 (August 7, 1995) 46.
18. Bruce R. Reichenbach and V. Elving Anderson, *On Behalf of God: A Christian Ethic for Biology* (Grand Rapids: William B. Eerdmans, 1995) 269.
19. Cited by L. Wright, "Double Mystery," 62.
20. Robert Wright, *The Moral Animal* (New York: Pantheon, 1994) 4–5.
21. Edward O. Wilson, *On Human Nature* (Cambridge: Harvard University Press, 1978) 71.
22. R. Wright, *Moral Animal*, 7: see: 45, 339. Although evolutionary psychology contradicts pluralism, it reinforces another characteristic of deconstructionist postmodernism, namely, the rather cynical belief that all human beliefs are driven by power interests. According to natural selection, only the fittest survive and all human culture, even human beliefs about truth, feed the fittest. "This belief helps nourish a central strand of the postmodern condition: a powerful inability to take things seriously." Ibid., 325. One can only ask of course: what genetic advantage is there to holding the postmodern-Darwinian belief that all belief is subject to the power interests of the fittest?
23. As "the chicken is only an egg's way of making another egg . . . [so] the organism is only DNA's way of making more DNA." Edward O. Wilson, *Sociobiology: The New Synthesis* (Cambridge MA: Harvard University Press, 1975) 3. Reichenbach and Anderson question the logic of Wilson's chicken- and egg-priority. "But why, one might ask, make the DNA more basic than the organism? Cannot we say that as the egg is the only way a chicken can make more chickens, so DNA is the only way an organism can make more organisms?" *On Behalf of God*, 271. Lewontin and colleagues object to the Wilson and Dawkins claim that the gene is logically prior to the individual and, correspondingly, "the individual to society, and equally explicitly on a set of transferred economic concepts . . . cost-benefit analysis, investment-opportunity costs, game theory, system engineering and communication, and the like are all unabashadly transferred into the natural domain." *Not In Our Genes*, 59–60.
24. Richard Dawkins, *The Selfish Gene* (New York: Oxford University Press, 1976, 1989) 200.
25. Wilson, *On Human Nature*, 167.
26. Michael Ruse puts it this way: "I argue that Darwinian factors inform and infuse the whole of human experience, most particularly our cultural dimension. . . . Human culture, meaning human thought and action, is informed and structured by biological factors. Natural selection and adaptive advantage reach through to the very core of our being." *Taking Darwin Seriously* (Oxford: Basil Blackwell, 1986) 140, 147.
27. Michael Ruse and Edward O. Wilson, "The Evolution of Ethics," *New Scientist*, 108:1478 (17 October 1985) 50–52.
28. Robert Wright, "Our Cheating Hearts," *Time*, 144:7 (August 15, 1994) 44–52.
29. Robert Wright, "The Biology of Violence," *The New Yorker*, 71:3 (March 13, 1995) 68–77.
30. R. Wright, *Moral Animal*, 146.
31. R. Wright, *Moral Animal*, 325.
32. R. Wright, *Moral Animal*, 147.
33. Evolutionary philosopher Michael Ruse, when trying to avoid committing the naturalistic fallacy, distinguishes selfishness-altruism at the genetic level from selfishness-altruism at the level of the human organism. "There is no implication that evolutionary altruism (working together for biological payoff) is inevitably

associated with moral altruism (*altruism* in the original literal sense, implying a conscious being helping others because it is right and proper to do so)." "Evolutionary Ethics: A Defense," in Rolston, *Biology, Ethics, and the Origins of Life*, 95. Although the latter is not identical with the former, Ruse believes that we cooperate at the cultural level because evolutionary biology has filled us with thoughts about right and wrong in order to achieve biologically determined ends. "The evolutionist's case is that ethics is a collective illusion of the human race, fashioned and maintained by natural selection in order to promote individual reproduction." Ibid., 101. Geneticist Francisco J. Ayala partially disagrees. "The sociobiologists' argument is that human ethical norms are sociocultural correlates of behaviors fostered by biological evolution. I argue that such proposals are misguided and do not escape the naturalistic fallacy. Perhaps it is true that both natural selection and moral norms sometimes coincide on the same behavior. The two are consistent. But this isomorphism between the behaviors promoted by natural selection and those sanctioned by moral norms exists only with respect to the consequences of the behaviors; the underlying causations are completely disparate." "The Difference of Being Human: Ethical Behavior as an Evolutionary Byproduct," Ibid., 118; see: 131–134.

34. R. Wright, *Moral Animal*, 31; see: 330–331.

35. R. Wright, *Moral Animal*, 10.

36. R. Wright, *Moral Animal*, 12–13. Wright operates with a naturalistic version of original sin. "The enemy of justice and decency does indeed lie in our genes. If ... I seem to depart from the public-relations strategy practiced by some Darwinists, and stress the bad in human nature more than the good, it is because I think we are more in danger of underestimating the enemy than overestimating it." Ibid., 151.

37. Jacques Monod emphasizes that science relies upon objective knowledge, that objective knowledge is the only way to truth, and that objective knowledge necessarily denies the existence of purpose in the phenomena it studies. Nevertheless, the teleonomic character of living organisms leads them to project and pursue purpose. *Chance and Necessity: An Essay on the Natural Philosophy of Modern Biology* (New York: Alfred A. Knopf, 1971) 21–22; 165. Random chance is what explains mutations and genetic variation in the history of evolution, not purpose. What happens to DNA is first by chance; and then, via natural selection and reproduction, it becomes necessity. Chance first, then necessity. So, purpose has evolved from what was previously without purpose. Monod then kicks this scientific observation up into a philosophical principle and argues for an atheistic naturalism, denying any divine purpose in the evolutionary advance because there is no foundational purpose at all. Biochemist and theologian Arthur Peacocke responds to Monod, agreeing that randomness and chance are as fundamental as natural law to the evolutionary advance. But Peacocke denies that this counts as evidence against he existence of a creator God. Rather, God uses chance to actualize multiple possibilities. "It is as if chance is the search radar of God, sweeping through all the possible targets available to its probing." *Creation and the World of Science* (Oxford: Clarendon Press, 1979) 95; see: *God and the New Biology* (San Francisco: Harper, 1986) 62.

38. Charles Darwin, *The Origin of the Species* (New York: Penguin, 1968) 263; cited by R. Wright, *Moral Animal*, 24.

39. R. Wright, *Moral Animal*, 162.

40. R. Wright, *Moral Animal*, 162–163.

41. R. Wright, *Moral Animal*, 35.

42. R. Wright, *Moral Animal*, 44.

43. R. Wright, *Moral Animal*, 148.

44. R. Wright, *Moral Animal*, 88.

45. R. Wright, *Moral Animal*, 37.

46. R. Wright, *Moral Animal*, 336.

47. R. Wright, *Moral Animal*, 37.

48. R. Wright, *Moral Animal*, 348.

49. R. Wright, *Moral Animal*, 349.

50. R. Wright, *Moral Animal*, 368; see: 13, 151.

51. R. Wright, *Moral Animal*, 377.

52. Lewontin, *Not In Our Genes*, 237. "Sociobiology is yet another attempt to put a natural scientific foundation under Adam Smith. It combines vulgar Mendelism, vulgar Darwinism, and vulgar reductionsism in the service of the status quo." Ibid., 264.

53. Philip Kitcher, *Vaulting Ambition: Sociobiology and the Quest for Human Nature* (Cambridge: MIT Press, 1990) 24.

54. Kitcher, *Vaulting Ambition*, 407.

55. See: Holmes Rolston III, *Genes, Genesis, and God: Beyond Selfishness to Shared Values* (New York: Columbia University Press, 1997).

56. Gilkey, "Biology and Theology on Human Nature," in Rolston, *Biology, Ethics, and the Origins of Life*, 185. Lewontin raises a similar critique regarding the sociobiologist's use of metaphor, calling it the error of reification. *Not In Our Genes*, 248–249.

57. Stanton L. Jones, "My Genes Made Me Do It," *Christianity Today*, 39:5 (April 24, 1995) 17.

58. Peacocke, *God and the New Biology*, 110–111.

59. Peacocke, *God and the New Biology*, 111. See: Peacocke, *Theology for a Scientific Age* (Minneapolis: Fortress Press, 1993) 226–227.

60. Michael Ruse wants sociobiology to prove its worth as a science. "If it can provide new insights for students of human behaviour, then no defence will be necessary," he writes. "If, like so many other bright ideas, it fails, then no defence will be sufficient." *Sociobiology: Sense or Nonsense?* (2nd ed. Boston: D. Reidel, 1985) xi.

61. Evelyne Shuster, "Determinism and Reductionism: A Greater Threat Because of the Human Genome Project?" *Gene Mapping*, 115.

62. Shuster, "Determinism and Reductionism," 116–117.

63. "DNA is quasimechanical; living organisms need not be." John Polkinghorne, *The Faith of a Physicist* (Princeton: Princeton University Press, 1994) 29.

64. Evelyn Fox Keller locates the gene myth not at the intersection between science and culture but rather by identifying science itself as culture, and as a cultural force with a vested power interest. She points to the concept of "genetic disease" as a social construction, an idea now on an imperialistic trajectory that represents the advancing influence of the field of molecular biology. The hidden contrast to the genetic disease would be the genetic norm, and because the gene myth teaches that our genome will teach us finally "what it means to be human," it follows that the field of molecular biology will become the cultural force legislating our concept of normality. "The nature of normality is allowed silently to elude the gaze of genetic scrutiny—and thereby tacitly to evade its determinist grip." "Nature, Nurture, and the Human Genome Project," *Code of Codes*, 298. Fox Keller's complaint is that this definiton of normality is merely negative—that is, a genome is normal if it has no disease alleles. The problem here is twofold. First, who gave the field of molecular biology the right to define normality for our culture? Second, Fox Keller implies that definitions of normality should include something positive rather than merely identify the lack of disease.

65. Nelkin and Lindee, *DNA Mystique*, 2.
66. Nelkin and Lindee, *DNA Mystique*, 16.
67. Nelkin and Lindee, *DNA Mystique*, 40–41.
68. Nelkin and Lindee, *DNA Mystique*, 55.
69. Nelkin and Lindee, *DNA Mystique*, 195–196.
70. Jeremy Gerard, "CBS Gives Rooney a 3-Month Suspension for Remarks," *New York Times* (February, 9 1990), cited in Nelkin, *DNA Mystique*, 171.
71. Nelkin and Lindee, *DNA Mystique*, 176.
72. Nelkin and Lindee, *DNA Mystique*, 181.
73. Nelkin and Lindee, *DNA Mystique*, 191.
74. Lewontin, Rose, and Kamin argue that "biological determinism has stood firm in claiming that occupying leadership roles in public, political, and cultural life goes with being male as much as having a penis, testicles, and facial hair." *Not In Our Genes*, 133.
75. Nelkin and Lindee, *DNA Mystique*, 196.
76. Hubbard, *Exploding the Gene Myth*, 5,6,8. Similarly, John Maddox contends that the "Strong Genetic Principle, that every aspect of the human condition is predetermined by the genes . . . is a fallacy." "Has Nature Overwhelmed Nurture?" 107.
77. Polkinghorne, *The Faith of a Physicist*, 82.
78. Polkinghorne, *The Faith of a Physicist*, 83.

Chapter 3
The Crime Gene, Stigma, and Original Sin

1. 354 Southeastern Reporter, 2nd Series, 124 (Georgia 1987). Free will is important to the courts. A conviction requires on the part of the accused *mens rea* (a guilty mind) plus criminal intent, both of which presuppose a free will. This is closely tied to the four reasons for sentencing: (1) retribution or punishment; (2) incapacitation to prevent further threats to society; (3) deterrence from commiting future crimes; and (4) rehabilitation or reform.
2. Mrs. Caldwell was released from prison and her family given responsibility to find an appropriate care facility for her.
3. For information: Huntington's Disease Society of America, Georgia Chapter, P.O. Box 245, Decatur, Georgia 30031.
4. Daughter Susan Caldwell is in a forgiving mood. "Part of the reason I was able to forgive my mother is because in her right mind I know she's not a cold-blooded killer," she told reporters. Susan added that she herself is a carrier of the Huntington's gene but does not herself have the disease. "She's Not a Cold-Blooded Killer," *Atlanta Journal Constitution* (September 28, 1994) A:1. Susan also said that "Huntington's disease was one of many factors in the crime my mother committed," others including the brain tumor. "A Complex Case," *Atlanta Journal Constitution* (October 4, 1994) A:18.
5. Quotation released by Associated Press (September 29, 1994).
6. "Woman with Brain Disorder Acquitted in Slaying of Son," *The Palm Beach Post* (September 24, 1994) Final Edition, A:12.
7. This discussion takes up the "unfinished business" material in chapter 10 of my earlier book, *Sin: Radical Evil in Soul and Society* (Grand Rapids: Wm. B. Eerdmans, 1994).
8. H.G. Brunner, M.R. Nelen, P. van Zandvoort, N.G.G.M. Abeling, A.H. van Gennip, E.C. Wolters, M.A. Kulper, H.H. Ropers, and B.A. van Oost, "X-Linked Borderline Mental Retardation with Prominent Behavioral Disturbance:

Phenotype, Genetic Localization, and Evidence for Disturbed Monoamine Metabolism," *American Journal of Human Genetics*, 52 (1993) 1032–1039.

9. Maureen P. Coffey, "The Genetic Defense: Explanation or Excuse?" *William and Mary Law Review*, 35:353 (1993) 395–396; italics in original.

10. "The Genetic Defense," 398. Coffey emphasizes the insanity defense. However, the courts can distinguish between insanity and other forms of diminished capacity. Children under a certain age, for example, are deemed as a matter of law incapable of committing a felony. Genetic predispositions to violence could be considered as diminished capacity to be held legally accountable for otherwise criminal behavior.

11. Patricia A. Jacobs, Muriel Brunton, Marie M. Melville, R.P. Brittain, W.F. McClemont, "Aggressive Behavior, Mental Subnormality and the XYY Male," *Nature*, 208 (December 25, 1965) 1351–1352.

12. A. Freyne and A. O'Conner, "XYY Genotype and Crime: 2 Cases," *Medicine, Science, Law*, 32:3 (1992) 261. See: Susan Mahler, "Genetic Predisposition and the Courts: Legal Fact or Science Fiction?" unpublished paper, 1994.

13. See: Jonathan Beckwith and Jonathan King, "The XYY Syndrome: A Dangerous Myth," *New Scientist*, (November 1974) 474–476.

14. Stephen Jay Gould, *The Mismeasure of Man* (New York: W.W. Norton, 1981) 145.

15. David Suzuki and Peter Knudtson, *Genethics: The Clash Between the New Genetics and Human Values* (Cambridge: Harvard University Press, 1990) 132.

16. Suzuki and Knudtson, *Genethics*, 138.

17. Suzuki and Knudtson, *Genethics*, 139.

18. Bruce R. Reichenbach and V. Elving Anderson, *On Behalf of God: A Christian Ethic for Biology* (Grand Rapids MI: Wm. B. Eerdmans, 1995) 199.

19. Patricia A. King, "The Past as Prologue: Race, Class, and Gene Discrimination," *Gene Mapping*, 102.

20. The frequently unnoted context is significant here. In 1992 then Secretary of Health and Human Services, Louis W. Sullivan, a black physician, conceived of a federal funding plan to help African Americans. He noted the prevalence of violent crimes in the black community, where the homicide rate is five times that for whites. His proposed program was to be "psycho-social" and examine a wide array of phenomena such as child abuse, drug addiction, and other potential factors in crime. The biological research component amounted to only 5 percent of the proposed budget. The assumption here was clearly that crime is socially linked to poverty and drug traffic. There was no attempt to reduce crime to genes. Yet the public atmosphere was later poisoned by Frederick K. Goodwin, then directing the Alcohol, Drug Abuse and Mental Health Administration and now heading the National Institute of Mental Health, who cited research on monkey violence and sexuality and then commented, "Maybe it isn't just the careless use of the word when people call certain areas of certain cities 'jungles.'" Civil rights leaders were fuming on the eve of the ill-fated University of Maryland conference. See: John Horgan, "Genes and Crime," *Scientific American*, 268:2 (February 1993), 24–29.

21. The brochure inviting participants included the following: "genetic research holds out the prospect of identifying individuals who may be predisposed to certain kinds of criminal conduct, of isolating environmental features which trigger those predispositions, and of treating some predispositions with drugs or unobtrusive therapies." Some readers of this brochure interpreted it to be racist.

22. Ruth Hubbard and Elijah Wald, "Responsible Science," *GeneWatch*, 8:4 (November 1992) 3.

23. Paul Billings, "Academic Freedom," *GeneWatch*, 8:4 (November 1992) 2. On September 3, 1993 the NIH Grant Appeals Board ruled that the decision to cancel the University of Maryland grant was based on spurious reasoning; and it reapproved the funding for the program to be rescheduled. See: "NIH Told to Reconsider Crime Meeting," *Science*, 262:5130 (October 1, 1993) 23.

24. Troy Duster, "Genetics, Race, and Crime: Recurring Seduction to a False Precision," *DNA on Trial: Genetic Information and Criminal Justice*, ed. by Paul Billings (Plainview, NY: Cold Spring Harbor Press, 1992) 132.

25. David Wasserman, "Science and Social Harm: Genetic Research into Crime and Violence," *Philosophy and Public Policy*, School of Public Affairs, University of Maryland, 15:1 (Winter 1995) 14–19, italics in original.

26. James Q. Wilson and Richard J. Herrnstein, *Crime and Human Nature* (New York: Touchstone, 1985) 66.

27. Wilson and Herrnstein, *Crime and Human Nature*, 508.

28. Wilson and Herrnstein, *Crime and Human Nature*, 69.

29. Troy Duster, *Backdoor to Eugenics* (New York: Routledge, 1990) vii.

30. Wilson and Herrnstein, *Crime and Human Nature*, 356.

31. Wilson and Herrnstein, *Crime and Human Nature*, 361.

32. Do genes play a role in alcoholism in women too? Lindon J. Eaves and colleagues conclude a study saying, "In women, genetic factors play a major etiologic role in alcoholism." See: Kenneth S. Kendler, Andrew C. Heath, Michael C. Neale, Ronald C. Kessler, and Lindon J. Eaves, "A Population-Based Twin Study of Alcoholism in Women," *JAMA*, 268:14 (October 14, 1992) 1881.

33. C. Robert Cloninger, "Neurogenetic Adaptive Mechanisms in Alcoholism," *Science*, 236 (April, 24 1987) 413. See Jerry E. Bishop and Michael Waldholz, *Genome* (New York: Simon and Schuster, 1990) 261.

34. At the turn of the twentieth century, the widely read novelist Jack London postulated that there are two types of alcoholism, the inherited type and the learned type. In his autobiographical book of 1913, *John Barleycorn*, he placed himself in the latter category. The eugenics movement already growing in Jack London's day placed "drunkards" along with prostitutes, criminals, and the feebleminded on their list of people in the first category, arguing that they should not propagate further their own defective genetic inheritance.

35. *Baker v. State Bar of California*, 781 P.2d 1344 (Cal. 1989). In a similar case the following year, *In re Ewaniszyk*, 788 P.2d 690 (Cal.1990), the California State Bar disbarred a convicted attorney despite proof of alcoholism. The distinguishing factor was not the *possession* of genetic predisposition to chemical addiction but rather Baker's *ignorance* of his biological handicap.

36. Dorothy Nelkin, "The Jurisprudence of Genetics," *Vanderbilt Law Review*, 45:2 (March 1992) 331.

37. Richard J. Herrnstein and Charles Murray, *The Bell Curve: Intelligence and Class Structure in American Life* (New York: Free Press, 1994) 22–23.

38. Herrnstein and Murray, *Bell Curve*, 25.

39. Herrnstein and Murray, *Bell Curve*, 117.

40. Herrnstein and Murray, *Bell Curve*, 270.

41. Herrnstein and Murray, *Bell Curve*, 548.

42. Herrnstein and Murray, *Bell Curve*, 519.

43. Herrnstein and Murray, *Bell Curve*, 548.

44. Robert A. Pyne, "The Ethical Ring of *The Bell Curve*," Dialog, 35:1 (Winter 1996) 54.

45. Alan Ryan, "Apocalypse Now?" *The New York Review* (November 17, 1994) 8.

46. Ibid.

47. Herrnstein and Murray, *The Bell Curve*, 297. Anticipating that they might be accused of contributing to racial prejudice against black people, Herrnstein and Murray appeal to the principle of individualism. Individuals are to be assessed according to their individual intelligence, not their ethnic average. "Many blacks would continue to be smarter than many whites. Ethnic differences would continue to be differences in means and distributions; they would continue to be useless, for all practical purposes, when assessing individuals. . . . *we cannot think of a legitimate argument why any encounter between individual whites and blacks need be affected by the knowledge that an aggregate ethnic difference in measured intelligence is genetic instead of environmental.*" Ibid., 312–313, italics in original.

48. "Genome Research Risks Abuse, Panel Warns," *Nature*, 378 (December 7, 1995) 529.

49. Lori B. Andrews and Dorothy Nelkin, "The Bell Curve: A Statement," *Science*, 271 (5 January 1996) 13–14.

50. Thomas Aquinas, *Summa Theologica*, II/1, q.84, a.4.

51. See: Augsburg Confession, Article II, in *The Book of Concord*, ed. by Theodore G. Tappert (Minneapolis: Fortress, 1959) 29, and John Calvin, *Institutes of the Christian Religion*, tr. by F.L. Battles, Vol. XX of *The Library of Christian Classics* (Louisville: Westminster/John Knox Press, 1960) 251.

52. *Book of Concord*, 510.

53. Calvin, *Institutes*, 151.

54. Speaking of evolutionary theory in light of the principle of maximum reproduction—that is, the selfish gene—William Irons notes how this fits well "with the Christian belief in original sin . . . part of the power of the belief in original sin stems from the fact that it portrays human beings as we actually experience them, that is, as having a potential for moral behavior combined with corruptibility." "How Did Morality Evolve?" *Zygon*, 26 (March 1991) 52.

55. Dorothy Nelkin and Laurence Tancredi, *Dangerous Diagnostics: The Social Power of Biological Information*. (New York: Basic Books, 1989) 12.

56. On the teaching of the Augustinian tradition that sin is universal though not necessary see Reinhold Niebuhr, *The Nature and Destiny of Man*, 2 Volumes (New York: Charles Scribner's Sons, 1941) 1:242. Robert John Russell helpfully delineates the Manichaean, Pelagian, and Augustinian systems. He has introduced the term "universal contingent" to describe sin as inevitable but not necessary. See his "The Thermodynamics of Natural Evil," *CTNS Bulletin*, 10:2 (Spring 1990) 20–35.

57. Augustine, "On the Merits and Remission of Sins" (*De Peccatorum Meritis et Remissione, et de Baptismo Parvulorum*) I:16.

58. *Book of Concord*, 514.

59. *Book of Concord*, 510.

60. Roger Haight, "Sin and Grace," *Systematic Theology: Roman Catholic Perspectives,* ed. by Francis Schüssler Fiorenza and John Galvin, 2 Volumes (Minneapolis: Fortress Press, 1991) II:85.

61. Paul Tillich, *Systematic Theology,* 3 Volumes (Chicago: University of Chicago Press, 1951–1963) II:46–47.

62. Marjorie Hewitt Suchocki, *The Fall to Violence: Original Sin in Relational Theology* (New York: Continuum, 1994) 12, 23.

63. Suchocki, *Fall to Violence*, 101, 104.

64. Suchocki, *Fall to Violence*, 118.

65. Suchocki, *Fall to Violence*, 126.

66. Suchocki, *Fall to Violence*, 129.

67. Suchocki, *Fall to Violence*, 132.

68. Suchocki, *Fall to Violence*, 145.

Chapter 4
The So-called "Gay Gene" and Scientized Morality

1. Dean H. Hamer, Stella Hu, Victoria L. Magnuson, Nan Hu, and Angela M.L. Pattatucci, "A Linkage Between DNA Markers on the X Chromosome and Male Sexual Orientation," *Science*, 261:5119 (July 16, 1993) 321–327.

2. William Henry III based on reports by Ellen Germain and Alice Park, "Born Gay?" *Time*, 142:4 (July 26, 1993) 36–39.

3. The question of whether homosexuality is sinful is a matter of dispute. We are not able to settle the dispute here. Rather, we are mapping different directions the ethical logic might take us. On one side of the dispute, however, we note that Christian ethics has traditionally relied upon the Levitical Holiness Code which prescribes the death penalty for homosexual acts (Lev. 18:22; 20:13) and that New Testament passages denounce homosexual relationships as idolatrous (Rom. 1:26–27) and indicate that certain kinds of same sex practices preclude entry into the kingdom of God (1 Cor. 6:9–10; 1 Tim. 1:9–10). We also note how today the Federation of Parents and Friends of Lesbians and Gays (FLAG) answers "no" to the question of their pamphlet title, "Is Homosexuality a Sin?" published in 1992 by P-FLAG, P.O. Box 27605, Washington DC 20038-7605. The pamphlet quotes former Lutheran Bishop Stanley E. Olson saying that "Diversity is beautiful in creation. How we live our lives in either affirming or destructive ways is God's concern, but being either homosexually oriented or heterosexually oriented is neither a divine plus or minus." Ibid., 9. Ethicist Karen Lebacqz associates sin not with homosexuality but with "homophobia, gay-bashing, discriminatory legislation toward lesbians and gays, refusal to include lesbian/gay bisexual people into our churches and communities." Ibid., 11. James B. Nelson identifies four distinct positions within Christian ethics regarding homosexuality: (1) the rejecting-punitive position; (2) the rejecting-nonpunitive position; (3) the qualified acceptance position; and (4) the full acceptance position. "Ethics" in *The Westminster Dictionary of Christian Ethics*, ed. by James F. Childress and John Macquarrie (Louisville: Westminster/John Knox Press, 1967, 1986) 273.

4. Simon A. LeVay, "A Difference in Hypothalamic Structure Between Heterosexual and Homosexual Men," *Science*, 253 (1991) 1034–1037; *The Sexual Brain* (Cambridge: M.I.T. Press, 1993). The LeVay study has garnered numerous criticisms, most frequently the objection to the fact that many of the corpses examined were persons who had died of AIDS. AIDS victims pose a research difficulty because they frequently have decreased testosterone levels. Some of the therapeutic drugs used with AIDS patients themselves are known to lower testosterone. With testosterone levels in flux, LeVay may not have gotten accurate readings. See: William Byne, "Science and Belief: Psychobiological Research on Sexual Orientation," in *Sex, Cells, and Same-Sex Desire*, ed. by John P. DeCecco and David Allen Parker (New York: Harrington Park Press, 1995) 329.

5. Dean Hamer with Peter Copeland, *The Science of Desire* (New York: Simon and Schuster, 1994) 138–139.

6. Hamer and Copeland, *Science of Desire*, 133.

7. Hamer and Copeland, *Science of Desire*, 134.

8. According to the Kinsey scale, four factors enter into the definition: (1) self-identification as either heterosexual, bisexual, or homosexual; (2) attraction to the same sex, opposite sex, or both; (3) sexual fantasies about one sex or the other or both; and (4) actual sexual behavior with men, women, or both. With 0 and 6 at the extreme ends of the continuum, someone at 3 would be relatively equal in bisexuality. Hamer dealt with no one self-identified at 3. *Science of Desire*, 66, 89. One of the weaknesses of the Kinsey scale is that it presupposes a bipolar

opposition between heterosexuality and homosexuality. If, for example, one wanted to measure intensity of sexual feeling or activity and a heterosexual at the 0 end was equal in intensity to a homosexual at the 6 end, and perhaps a bisexual at 3 showed relatively little sexual interest, the Kinsey scale would be useless if not misleading.

9. We know that in rare instances a man may be XX and woman XY. Closer inspection shows that in cases of XX men a small piece of a Y chromsome has been transferred to one of the Xs. And in cases of XY women a small but essential portion of the Y chromosome is lacking. Gender, it turns out, is due to TDF. Hamer and Copeland, *Science of Desire*, 153.

10. Hamer and Copeland, *Science of Desire*, 153.

11. Hamer writes, "If our most important sexual organ is the brain, perhaps the Xq28 locus influences sexual orientation by altering the brain's structure or chemical activity." Hamer and Copeland, *Science of Desire*, 160. See: Simon LeVay and Dean H. Hamer, "Evidence for a Biological Influence in Male Homosexuality," *Scientific American*, 272:5 (May 1994) 44–49.

12. Hamer and Copeland, *Science of Desire*, 150. This brings a scientific word to bear on the postmodern social constructivist interpretation of homosexuality. Michael Foucault has argued that homosexuality as we know it today is a social construction of the nineteenth century. He actually dates it at 1870. The transition here is from homosexual acts such as sodomy to the notion that homosexuality is constitutional, that it belongs at the core of a person's being. It became accepted that being gay is "consubstantial with him, less a habitual sin than as a singular nature . . . a kind of interior androgyny, a hermaphrodism of the soul. The sodomite had been a temporary aberration; the homosexual was now a species." *The History of Sexuality*, 3 Volumes (New York: Vintage Books, 1978–1986) I:43.

13. William Byne, "The Biological Evidence Challenged," *Scientific American*, 274:5 (May 1994) 50–55.

14. Reported by Daniel J. Kevles, "The X Factor," *The New Yorker*, LXXI:6 (April 3, 1995) 89.

15. Rainer Herrn, "On the History of Biological Theories of Homosexuality," *Sex, Cells, and Same-Sex Desire*, 40.

16. Herrn, *Sex, Cells, and Same-Sex Desire*, 49.

17. John P. DeCecco and David Allen Parker, "Sexual Expression: A Global Perspective," in *Sex, Cells, and Same-Sex Desire*, 428.

18. Hamer and Copeland, *Science of Desire*, 177.

19. Hamer and Copeland, *Science of Desire*, 213.

20. Much of the following discussion appeared initially in my article, "On the Gay Gene: Back to Original Sin Again?" *Dialog*, 33:1 (Winter 1994) 30–38.

21. The vocabulary of Janet E. Halley is helpful in sorting out the possible positions. *Progay* essentialism holds that because homosexual orientation is fixed, immutable, and definitional, it should be protected from discrimination. *Progay constructivism* holds that all forms of sexual orientation are mutable, affected by free choice, and that no social policy should impede these variations. *Antigay essentialism* holds that homosexual orientation is not only immutable but also abnormal, bad, or sick and that its manifestation should be impeded and homosexual individuals cured. *Antigay constructivism* holds that sexual orientation is mutable, the result of choice, and that discrimination should be employed to convert gay men and lesbian women to heterosexuality. Halley, a Stanford Law professor and herself a bisexual, argues that the progay essentialist argument leaves out the interests of others whose sexuality is mutable or the result of free choice. She recommends an alliance of essentialists and constructivists against

discrimination. "Sexual Orientation and the Politics of Biology: A Critique of the Argument from Immutability," *Stanford Law Review*, 46:3 (February 1994) 503–568.

22. Karen De Witt, "Quayle Contends Homosexuality is a Matter of Choice, Not Biology," *New York Times* (September 14, 1992) A17.

23. See: David J. Jefferson, "Studying the Biology of Sexual Orientation Has Political Fallout," *Wall Street Journal*, CXXIX:30 (August 12, 1993) A4.

24. Cited by Kim Painter, "Key Evidence: More Maternal Kin are Gay," *USA Today*, (July 16–18, 1993) A2.

25. An extreme form of this position might be biological innocence with a morally good gay gene on the grounds that if God or nature gives it, then it must be good. Some might want to argue that the gay gene is a "gift" of creation and, therefore, homosexual behavior is morally justified. But this would be a clear example of the naturalistic fallacy: the false assumption that what is natural is good. It violates Hume's law prohibiting us drawing an *ought* from an *is*. William Irons, writing on the evolutionary basis of morality, says, "we are forced to recognize that we cannot judge behaviors moral or immoral by evaluating their naturalness. Evidence indicating that a particular behavior is natural tells us nothing about its morality. Natural behaviors may be moral, immoral, or morally neutral." "How Did Morality Evolve?" *Zygon* 26:1 (March 1991) 52.

26. "If homosexuality were found to be an immutable trait like skin color, then laws criminalizing homosexual sex might be overturned," suggests David J. Jefferson. A Hawaii supreme court judge has noted that a biological basis for homosexuality could tip the scales toward legalization of same-sex marriages. "Studying the Biology of Sexual Orientation has Political Fallout," *The Wall Street Journal*, XCCIX:30 (August 12, 1993) A1.

27. Hamer and Copeland, *Science of Desire*, 216.

28. Kevles, "The X Factor," 89–90.

29. Augustine, "The Spirit and the Letter" 20.

30. Wolfhart Pannenberg, "You Shall Not Lie with a Male," *Lutheran Forum*, 30:1 (February 1996) 28.

31. Pannenberg, "You Shall Not Lie with a Male," 29.

32. "An Open Letter from the Bishops of the Evangelical Lutheran Church in America" (March 22, 1996), obtainable from ELCA, 8765 Higgins Road, Chicago, IL 60631.

Chapter 5
Should We Patent God's Creation?

1. This chapter is a revised version of my "Theology Update" titled "Should We Patent God's Creation?" in *Dialog* 35:2 (Spring 1996) 117–132.

2. At this writing the Patent and Trademark Office is planning to hold hearings on philosophical issues: Should the United States patent "a complete genome of an organism" or "human genome fragments," and would this "inhibit rather than promote the advancement of the biotechnology arts"? *Science*, 270 (24 November 1995) 1287.

3. "Religious Leaders Prepare to Fight Patents on Genes," *New York Times*, CXLIV:50,060, National Edition (May 13, 1995) 1.

4. "This is not the first time Rifkin has joined religious leaders to mount a crusade against genetic engineering," reports *Science* magazine. "In June 1983, he persuaded fifty religious leaders to sign a resolution opposed to efforts to engineer specific genetic traits into the germline of the human species. Four years later . . . Rifkin organized a religious coalition opposed to the patenting of genetically

altered animals." Regarding the May 18, 1995 event, however, the impetus seems to have come from the United Methodist Church. Richard Stone, "Religious Leaders Oppose Patenting Genes and Animals," *Science*, 268:5214 (26 May 1995) 1126.

5. Already appearing in the "United Methodist Church Genetic Science Task Force Report to the 1992 General Conference" is the assertion that "the patenting of life forms is a crucial issue." This is accompanied by a theological argument denying certain patents on the grounds that life is sacred and that God is the only proper owner of life. "[O]ur understanding is of the sanctity of God's creation and God's ownership of life. Therefore, exclusive ownership rights of genes as a means of making genetic technologies accessible raises serious theological concerns. While patents on organisms themselves are opposed, process patents . . . can be supported" (117). General Board of Church and Society of the United Methodist Church, 100 Maryland Avenue, N. E., Washington, DC 20002.

6. The U.S. Patent and Trademark Office (PTO) grants to a patent holder the exclusive right to prevent others from making, using, or selling a process, manufactured machine, or composition of matter that he or she has invented for a limited period of time. Up until recently that limited period has been seventeen years. As of June 8, 1995, the patent term is now twenty years from the filing date.

7. Rifkin, *Algeny,* 252.

8. Ronald Cole-Turner makes this clear. "Rifkin's ideas are those of a vitalist, whereas most religious traditions in America today are theistic. A vitalist tends to see all life as sacred and thus as off-limits for alteration or ownership. Theists, however, believe that only God is sacred. Everything else is God's creation, and although creation should be treated with respect, there is no metaphysical difference between DNA and other complex chemicals." *Science*, 270:5233 (October 6, 1995) 52.

9. Andrew Kimbrell, *The Human Body Shop* (San Francisco: Harper, 1993) 210–211.

10. Ibid.

11. As reported by Karen Bernstein, "Religious Move to Stop Gene Patents," *BioCentury*, 3:37 (May 15, 1995) A2.

12. In a statement issued separately on May 10, 1995, "An Introduction to Genes and Patents," Bishop William B. Friend, who along with ninety other Roman Catholic Bishops signed the Joint Appeal document, added a special concern not reflected in the document itself: the distinction between animal genes and human genes. Roman Catholic thinking focuses on human persons, not animals, and this requires a distinctive set of moral principles and theological teachings.

13. Philip Hefner, *The Human Factor* (Minneapolis: Fortress, 1993) 35–42.

14. Kimbrell, *The Human Body Shop*, 199, 202.

15. "Who Speaks for God?" *BioCentury*, 3:37 (May 15, 1993) A4.

16. Ibid.

17. Necessary here is human intervention that leads to something new. Title 35, Section 101, of the United States Code declares that patents may be obtained for new and useful processes, machines, articles of manufacture, and compositions of matter. In the Chakrabarty case to be discussed later in this chapter, the Supreme Court declared that "literally anything under the sun known to mankind" could be patented if it involves human intervention.

18. Cited by Christopher Anderson, "NIH Drops Bid for Gene Patents," *Science*, 263 (February 18, 1994) 909–910.

19. Venter and his colleague, William A. Haseltine, CEO of Human Genome Sciences, made the cover of *Business Week*, with an article by John Carey, "The Gene Kings" (May 8, 1995) 72–78.
20. Watson, "A Personal View of the Project," in *The Code of Codes*, 164–173.
21. Cited in L. Roberts, "NIH Gene Patents, Round Two," *Science*, 255 (1992) 912.
22. P. Zurer, "Critics Take Aim at NIH's Gene Patenting Strategy," *Chemical and Engineering News*, 70 (1992) 21–22.
23. C. Anderson, "To Patent a Naked Gene," *Nature*, 353 (1991) 485. Rebecca S. Eisenberg says that the argument against allowing NIH to patent Venter's sequences is not really that these sequences are useless; rather, it is that we do not yet know what they are good for. We should not be able to claim patent rights ahead of subsequent researchers who will attempt to figure this out. "Genes, Patents, and Product Development," *Science* 257 (August 14, 1992) 903–908.
24. 33 US 130 (1948); cited by Rebecca Eisenberg, "Genes, Patents, and Product Development," *Science*, 257 (August 14, 1992) 903–908.
25. Kate H. Murashige, "Intellectual Property and Genetic Testing," in *The Genetic Frontier: Ethics, Law, and Policy*, ed. by Mark S. Frankel and Albert Teich (Washington, DC: AAAS, 1994) 181–198. The actual language for patenting based on Article 1 of the U.S. Constitution uses the phrase "invention or discovery." Yet, discovery here should be interpreted as discovering something new, not as discovering something already existing in nature.
26. Kevles and Hood, *The Code of Codes*, 313, 314.
27. "HUGO Statement on Patenting of DNA Sequences," HUGO Americas, 7986–D Old Georgetown Road, Bethesda, MD 20814, USA. Report in *Human Genome News*, 6:6 (March–April 1995) 5. See: "The CRG Says No to Patenting Life Forms," by Stuart Newman and Nachama Wilker, for a similar position taken by the Council for Responsible Genetics, 5 Upland Road, Suite 3, Cambridge, MA 02140.
28. Kimbrell, *The Human Body Shop*, 190.
29. One of the most heated subcontroversies within the larger patenting controversy has to do with objections raised against U.S. government patenting of third-world cell lines by the Rural Advancement Foundation International (RAFI). RAFI blew the whistle on a U.S. Department of Commerce 1993 patent application for a cell line taken from Guaymi Indians in Panama's rain forest, the purpose of which was to aid medical research. They have also objected to a patent deal with the Hagahai tribe in Papua, New Guinea, even though the tribe would receive half the royalties. RAFI believes there is something ethically wrong with patenting life forms, including human genetic material. In its "Blue Mountain Declaration" RAFI states its oppostion to the conversion of "life forms, their molecules or parts into corporate property through patent monopolies ... [No one] should be able to hold patents on organs, cells, genes or proteins, whether naturally occurring, genetically altered or otherwise modified." See: Philip L. Bereano, "Genetic Patents," *Science*, 271:5245 (January 5, 1996) 14.
30. Zurer, "Critics take Aim," 22.
31. R. Thompson, "The Race to Map Our Genes," *Time*, 141 (1993) 57.
32. The fights that usually take place between scientists are less often over what should get patented but rather whose names should be included on the patent. In 1995 Gene Therapy Inc. (GTI) announced it had taken exclusive rights to a NIH patent on ex vivo gene therapy—wherein a therapeutic gene is inserted into cells that have been temporarily removed from the patient's body. The pulse of many geneticists shot up as fast as the company's stock. Only three names of

inventors appeared on the patent: French Anderson, Michael Blaese, and Steven Rosenberg. The disgruntled complainers said that one hundred researchers had worked on the NIH team that developed the therapy and, furthermore, that prior to the patent application others such as the Salk Institute had been using it. Rachel Nowak, "Patent Award Stirs a Controversy," *Science*, 267:5206 (March 31, 1995) 1899.

33. Rebecca S. Eisenberg, "Patenting the Human Genome," *Emory Law Journal*, 39:3 (Summer 1990) 721–745, 739.

34. Eisenberg cites Pub. L. No. 96–517, 94 Stat. 3015, 3019–29 [1980 (codified at 35 U.S.C. pars. 200–211 (1988)]. Ibid.

35. Jeff Lyon and Peter Gorner complain that this game of musical labs is played by shamelessly greedy biologists whose desire for pecuniary gain interferes with the process of seeking truth. *Altered Fates: Gene Therapy and the Retooling of Human Life* (New York: W.W. Norton, 1995). In his review of this book, Malcolm Gladwell rhetorically argues to the contrary. "Why is the rush by scientists to form companies . . . evidence of greed and shamelessness? How, exactly, does the pursuit of pecuniary gain interfere with the process of seeking the truth? Isn't the extraordinary access that gene scientists have to biological 'truth' the reason they are worth so much money? And wouldn't the real ethical lapse be for gene therapists, having conceived of technologies with vast and immediate therapeutic value, not try and bring them to market as quickly as possible?" "Rights to Life," *The New Yorker*, LXXI:36 (November 13, 1995) 121.

36. Eisenberg, "Patenting the Human Genome," 744.

37. Eisenberg, "Genes, Patents, and Product Development," 904. See her unpublished paper, "A Technology Policy Perspective on the NIH Gene Patenting Controversy," 1994.

38. 447 U.S. 308–309 (1980) cited by Eisenberg, "Patenting the Human Genome," 725. Kimbrell reports that Jeremy Rifkin and Ted Howard filed an amicus brief to support nonpatentability, arguing that a ruling supporting the patenting of a life form would in effect regard life as "less than life, as nothing but common chemicals." *The Human Body Shop*, 194. Kimbrell's assesment: "The complete failure by the Court to correctly assess the impacts of the Chakrabarty decision may go down as among the biggest judicial miscalculations in the Court's long history." (195).

39. Commissioner's Notice, *Official Gazette of the Patent and Trademark Office*, 1077 (April 21, 1987) 24. My reporting and analysis in this section is dependent on the helpful research of Rebecca Eisenberg.

40. Kimbrell is worried that this exemption of human beings from patenting may not be strong enough. "Unfortunately, there were several problems with the human exemption. For one, under the PTO's 1987 ruling, embryos and fetuses, human life-forms not presently covered under Thirteenth Amendment protection, are patentable, as are genetically engineered human tissues, cells, and genes. In fact, a genetically engineered human kidney, cornea, arm, or leg, or any other body part might well be patentable." *The Human Body Shop*, 199. An even more volatile set of accusations including that of "biocolonialism" aimed at first world scientists who patent cell lines discovered among indigenous peoples is repeatedly raised by the Rural Advancement Foundation International (RAFI). Alluding to scientific "vampires," this group accused medical anthropologist Carol Jenkins and her NIH collaborators for patent theft of genetic material belonging to the Hagahai tribe in New Guinea. An electronic press release claimed that an "indigenous person" had been patented. Defending herself, Jenkins reported that the Hagahai agreed to receive 50 percent of patent royalties. Stanford University Law professor and chair of the ethics committe of

the Human Genome Diversity Project told the press that "the patent doesn't patent a person. It doesn't even patent human genetic material. It's the cell line, a viral preparation derived from the cell line, and three different bioassays to see whether people are infected by this virus. And the idea that the U.S. government owns this person or his genetic material is absolute rubbish." Gary Taubes, "Scientists Attacked for 'Patenting' Pacific Tribe, *Science*, 270 (17 November 1995) 1112.

41. "Who Speaks for God?" A5.

Chapter 6
The Question of Germline Intervention

1. Karl Rahner, "Christology within an Evolutionary View," *Theological Investigations*, 22 Volumes (London: Darton, Longman & Todd, and New York: Crossroad, 1961–1988) V:168; see XXI:54.

2. Rahner, *Theological Investigations*, V:137–138.

3. Burke K. Zimmerman identifies "three strategies" for germline intervention: (1) screening and selection of early-stage embryos; (2) direct modification of the DNA of early-stage embryos coupled with IVF; and (3) genetic modification of gametes prior to conception. Although screening is not usually included in germline discussion, it, along with the other two strategies, would affect the future human gene pool. "Human Germ-Line Therapy: The Case for its Development and Use." *Journal of Medicine and Philosophy*, 16:6 (December 1991) 593–612, esp. 594–595. See also Gregory Fowler, Eric Juengst, and Burke K. Zimmerman, "Germ-Line Therapy and the Clinical Ethos of Medical Genetics," *Theoretical Medicine*, 10 (1989) 151–165.

4. This chapter builds upon material developed for a paper delivered at a conference held in Rome on biological evolution cosponsored by the Vatican Observatory and the Center for Theology and the Natural Sciences in June 1996 as well as material appearing in my article, "Playing God and Germline Intervention," *Journal of Medicine and Philosophy*, 20:4 (August 1995) 365–386.

5. Paul Ramsey writes, "Men ought not to play God before they learn to be men, and after they have learned to be men they will not play God." Paul Ramsey, *Fabricated Man: The Ethics of Genetic Control* (New Haven: Yale, 1970) 138. The question of playing God in genetic intervention is only one of many reasons for inviting theological attention into this field. M. Therese Lysaught formerly at the Park Ridge Center for the Study of Health, Faith, and Ethics writes, "a Christian theological analysis of the Human Genome Project and genetics needs to examine a host of questions in addition to the question of human intervention into nature—for example, questions of theodicy, of divine agency, of theological anthropology, of social justice, of the meaning of suffering within a Christian theological framework, of the meaning of Christian community, as well as methodological questions surrounding the science/religion dialogue." M. Therese Lysaught, "Map, Myth, or Medium of Redemption: How Do We Interpret the Human Genome Project," *Second Opinion*, 19:4 (April 1994) 83.

6. John C. Fletcher and W. French Anderson, "Germ-Line Gene Therapy: A New Stage of Debate," *Law, Medicine, and Health Care*, 20:1/2 (Spring/Summer 1992) 31.

7. W. French Anderson, for example, writes, "Somatic cell gene therapy for the treatment of severe disease is considered ethical because it can be supported by the fundamental moral principle of beneficience: It would relieve human suffering. . . . [But] enhancement engineering would threaten important human values in two ways: It could be medically hazardous in that the risks could exceed the

potential benefits and the procedure therefore cause harm. And it would be morally precarious in that . . . it could lead to an increase in inequality and discriminatory practices." "Genetics and Human Malleability," *Hastings Center Report*, 20:1 (January February, 1990) 23. David Suzuki and Peter Knudtson draw a sharp ethical distinction between somatic gene therapy—which can be seen as the equivalent of an organ-transplant operation that modifies a patient's phenotype without changing the genotype—and germline gene therapy—which modifies "cells belonging to lineages that are potentially immortal." David Suzuki and Peter Knudtson, *Genethics,* 183–184. Their position is clear: "Germ-cell therapy, without the consent of all members of society, ought to be explicitly forbidden." (335). However, because the technical distinction between these two is becoming more difficult to discern, some can say "the bright ethical line separating somatic and germline therapy has begun to erode." Kathleen Nolan, "How Do We Think About the Ethics of Human Germ-Line Genetic Therapy," *Journal of Medicine and Philosophy*, 16:6 (December 1991) 613. French Anderson, in a more recent work with John Fletcher, argues that the situation is changing. Whereas in the 1970s and 1980s there was a strong taboo against germline modification, in the 1990s that taboo is lifting. "Searches for cure and prevention of genetic disorders by germline therapy arise from principles of beneficence and nonmaleficence, which create imperatives to relieve and prevent basic causes of human suffering. It follows from this ethical imperative that society ought not to draw a moral line between intentional germline therapy and somatic cell therapy." Fletcher and Anderson, "Germ-Line Gene Therapy," 31.

8. World Council of Churches, *Manipulating Life: Ethical Issues in Genetic Engineering. Geneva: WorldCouncil of Churches*, 1982. A 1989 document reiterates this position more strongly by proposing "a ban on experiments involving genetic engineering of the human germline at the present time." WCC, "Biotechnology: Its Challenges to the Churches and the World." Unpublished. Geneva: World Council of Churches, 1989, 2. Eric T. Juengst argues that the arguments for a present ban on germline intervention are convincing, but he argues that the risks of genetic accidents—even multigenerational ones—can be overcome with new knowledge. Germline alteration ought not be proscribed simply on the grounds that enhancement engineering might magnify current social inequalities. He writes, "the social risks of enhancement engineering, like its clinical risks, still only provides contingent barriers to the technique. In a society structured to allow the realization of our moral commitment to social equality in the face of biological diversity—that is, for a society in which there was both open access to this technology and not particular social advantage to its use—these problems would show themselves to be the side issues they really are." Eric T. Juengst, "The NIH 'Points to Consider' and the Limits of Human Gene Therapy," *Human Gene Therapy*, 1 (1990) 431.

9. United Methodist Church Genetic Task Force Report to the 1992 General Conference, 121.

10. United Church of Christ, "The Church and Genetic Engineering," Pronouncement of the Seventeenth General Synod, Fort Worth, Texas 1989, 3. See: Ronald Cole-Turner, "Genetics and the Church," *Prism*, 6 (Spring 1991) 53–61; and *The New Genesis*, 70–79.

11. See: Paul Nelson, "Bioethics in the Lutheran Tradition," *Bioethics Yearbook*, Volume 1: *Theological Developments in Bioethics: 1988–1990* (Boston: Kluwer, 1991) 119–144. Reporting on recent developments in Scandanavian theology, Paul Nelson writes, "Biblically oriented theologians reject positive eugenics because human nature as created and willed by God rests upon a genetic foundation. Modifications aimed at making better humans would usurp the author-

ity of the divine creation and efface the distinction between the creature and creator. . . . At the same time, these theologians are not opposed to negative eugenics in the form of somatic cell gene therapy. . . . Germ cell therapy, on the other
hand, is subject to the same indictment the churches make of positive eugenics."
Volume 3: *Theological Developments in Bioethics: 1990–1992* (Boston: Kluwer,
1993) 161. Ethicist Ernlé W.D. Young puts his opposition to germline repair
this way: "a treatment that produces an inheritable change, and could therefore
perpetuate in future generations any mistake or unanticipated problems resulting from therapy, is fraught with risk." *Alpha and Omega: Ethics at the Frontiers of Life and Death* (New York: Addison-Wesley, 1989) 49.

12. Catholic Health Association of the United States, *Human Genetics: Ethical
Issues in Genetic Testing, Counseling, and Therapy.* St. Louis: The Catholic
Health Association of the United States, 1990, 19. Paulus Gregorius, Metropol-
itan Orthodox Bishop of Delhi, might agree. "[O]ne cannot see anything intrin-
sically forbidden or evil in gene therapy, whether somatic or germline." Paul
Gregorius, "Ethical Reflections on Human Gene Therapy," in Zbigniew
Bankowski, Alexander M. Capron, eds., *Genetics, Ethics, and Human Values.*
Proceedings of the Twenty-Forth CIOMS Round Table Conference. (Geneva:
CIOMS, 1991) 143–53.

13. See: Peter Meyer, "Biotechnology: History Shapes German Question," *Forum
for Applied Research and Public Policy,* 6 (Winter 1991) 92–97.

14. See: Duster, *Backdoor to Eugenics;* Hubbard and Wald, *Exploding the Gene
Myth,* esp. 24–25; and Rifkin, *Algeny,* 230–234.

15. John Harris, "Is Gene Therapy a Form of Eugenics?" *Bioethics,* 7:2/3 (April
1993) 184.

16. J. Robert Nelson, "Summary Reflection Statement" of the "Genetics, Religion
and Ethics Project" (1992), The Institute of Religion and Baylor College of
Medicine, The Texas Medical Center, P.O. Box 20569, Houston, Texas 77225.
For an analysis of the conference, see J. Robert Nelson, *On the New Frontiers
of Genetics and Religion* (Grand Rapids: William B. Eerdmans, 1994).

17. Eric T. Juengst, "Germ-Line Gene Therapy: Back to Basics," *Journal of Medicine and Philosophy,* 16:6 (December 1991) 589–590. See also Maurice A.M.
De Wachter, "Ethical Aspects of Human Germ-Line Therapy," *Bioethics,* 7:2/3
(April 1993) 166–177. Nelson A. Wivel and LeRoy Walters list four arguments
against germline modification: (1) it is an expensive intervention that would
affect relatively few patients; (2) alternative strategies for avoiding genetic disease exist, namely, somatic cell therapy; (3) the risks of multigenerational genetic
mistakes will never be eliminated, and these mistakes would be irreversible; and
(4) germline modification for therapy puts us on a slippery slope leading
inevitably to enhancement. They also list four arguments favoring germline
modification: (1) health professionals have a moral obligation to use the best
available methods in preventing or treating genetic disorders, and this may
include germline alterations; (2) the principle of respect for parental autonomy
should permit parents to use this technology to increase the likelihood of having
a healthy child; (3) it is more efficient than the repeated use of somatic cell therapy over successive generations; and (4) the prevailing ethic of science and medicine operates on the assumption that knowledge has intrinsic value, and this
means that promising areas of research should be pursued. "Germ-Line Gene
Modification and Disease Prevention: Some Medical and Ethical Perspectives,"
Science, 262:5133 (October 22, 1993) 533–538. Arthur L. Caplan believes HGI
scientists may have sold their research souls too soon by promising to refrain
from germline intervention just to appease the hysteria over potential eugenic
uses. There is no moral reason to refrain from eliminating a lethal gene from the

human population; and there is no slippery slope from germline therapy to eugenics. "It is simply a confusion to equate eugenics with any discussion of germline therapy." "If Gene Therapy Can Cure, What is the Disease?" in *Gene Mapping*, 139.

18. C. Thomas Caskey, like the CRG, believes that germline correction has little practical appeal while generating considerable ethical apprehension. Yet, he leaves the door open. "I would reserve one area for consideration of germline manipulation. . . . It is conceivable that at some point in the future genetic manipulation of an individual's germline may be undertaken to introduce or reintroduce disease resistance." C. Thomas Caskey, "DNA-Based Medicine: Prevention and Therapy," in *The Code of Codes,* 129. John A. Robertson takes a position that would oppose the CRG, saying that "these fears appear too speculative to justify denying use of a therapeutic technique that will protect more immediate generations of offspring." *Children of Choice: Freedom and the New Reproductive Technologies* (Princeton: Princeton University Press, 1994) 162.

19. See: Daniel J. Kevles, *In the Name of Eugenics* (Berkeley and Los Angeles: University of California Press, 1985).

20. See: Robert N. Proctor, *Racial Hygiene: Medicine Under the Nazis* (Cambridge: Harvard University Press, 1988).

21. This statement comes directly from the position paper. It fits appropriately with what one of the drafters, Richard C. Lewontin, elsewhere says critically about science and class interests: "'Science' is the ultimate legitimator of bourgeois ideology." Richard C. Lewontin, et.al., *Not In Our Genes* (New York: Pantheon, 1984) 31.

22. Bouma, *Christian Faith, Health, and Medical Practice*, 264.

23. "Why does the notion that medical technology might give some children an advantage elicit such a strong negative reaction?" asks Zimmerman. "Perhaps it is because the notion of fairness is well embedded in Western culture." He goes on to note that we already accept randomized differences between people and the inevitability that some individuals will excell over others. Then in support of germline enhancement he adds: "What about the positive side, of increasing the number of talented people. Wouldn't society be better off in the long run?" "Human Germ-Line Therapy," 606–607.

24. We must be clear that genetic prejudice would be a cultural or social phenomenon, not a scientific one. "It is society, not biology, that turns some genetic characteristics into liabilities," writes Roger L. Shinn, *Forced Options: Social Decisions for the 21st Century* (New York: Pilgrim Press, 2nd ed., 1985) 140. If our society is serious about the fairness or justice dimension here, we could institute a sort of "Affirmative Action" public policy in which the underprivileged classes would be given privileged access to germline enhancement technology.

25. Cole-Turner makes much of Jesus' healing ministry as a directive toward inspiring contemporary science and technology to continue healing and to think of this as continuing the divine work of redemption. *New Genesis*, 80–86.

26. Hardy Jones, "Genetic Endowment and Obligations to Future Generations," in *Responsibilities to Future Generations*, ed. by Ernest Partridge, (Buffalo: Prometheus Books, 1981) 249.

27. Robertson, *Children of Choice*, 162.

28. David Suzuki and Peter Knudtson promulgate a "genethic principle" that parallels the CRG: "While genetic manipulation of human somatic cells may lie in the realm of personal choice, tinkering with human germ cells does not. Germ-cell therapy, without the consent of all members of society, ought to be explicitly forbidden." *Genethics*, 163. The Suzuki and Knudtson position is obviously based upon a libertarian ethic so, to be more precise, they should be seeking the

consent of those individuals involved rather than the vague "all members of society." Philip Kitcher cautions against both somatic and germline therapy now due to unforeseen damage; yet, he would approve of both when more advanced technology reduces risk. He adds that "it would be folly to oppose the therapy on the grounds that the children have not consented to it." *The Lives to Come* (New York: Simon and Schuster, 1996) 123.

29. Wolfhart Pannenberg, *Ethics* (Louisville: Westminster/John Knox Press, 1981) 140.
30. Cole-Turner, *New Genesis*, 98.
31. Karl Rahner, *Foundations of Christian Faith* (New York: Seabury, 1978) 35.
32. Rahner, *Foundations*, 190.
33. Rahner, *Theological Investigations*, IX:211.
34. Roger Shinn's advice is salutary here. "I know of no way of drawing a line and saying: thus far, scientific direction and control is beneficial; beyond this line they become destructive manipulation. I think it more important to keep raising the question, to keep confronting the technological society with the issue." *Forced Options*, 142. Deborah Blake says it eloquently: "The risk of the nineties is the seduction of a technological fix. The challenge for the nineties is to find the moral courage necessary to guide and realize the promises made by this new genetics so that our moral wisdom is not outpaced by our technological cleverness." "Ethics of Possibility: Medical Biotechnology for the Nineties," *The Catholic World*, 234:1403 (September-October, 1991) 237. Such thinking leads Roger Shinn to favor somatic therapy while opposing germline intervention; but he still wants to "keep the window open a crack." *The New Genetics* (Wakefield, RI: Moyer Bell, 1996) 145.

Chapter 7
A Theology of Freedom

1. David Tracy, *The Analogical Imagination: Christian Theology and the Culture of Pluralism* (New York: Crossroad, 1981) 64.
2. Paul Tillich, *Systematic Theology*, 3 Volumes (Chicago: University of Chicago Press, 1951–1963) III:158. See his *Theology of Culture* (Oxford: Oxford University Press, 1959) Chapter 1.
3. Langdon Gilkey, *Society and the Sacred* (New York: Crossroad, 1991) x.
4. Marjorie Hewitt Suchocki, *The Fall to Violence* (New York: Crossroad, 1994) 132.
5. Suchocki, *The Fall to Violence*, 130.
6. Suchocki, *The Fall to Violence*, 131.
7. John Macquarrie, *In Search of Humanity: A Theological and Philosophical Approach* (New York: Crossroad, 1983) 12.
8. Macquarrie, *In Search of Humanity*, 14.
9. Although Macquarrie does not use Philip Hefner's term "created cocreator," the idea is present when he writes, "the human being has a share in creating a human essence. Humanity is not a finished product. The human race is coming to be, living in history, seeking a goal which would be a fuller being, both for individuals and for the whole race. At least within limits, this goal can be chosen and humanity can determine what it is to become." *In Search of Humanity*, 15.
10. Tillich, *Systematic Theology*, I:169. Hereinafter ST.
11. Tillich, ST, I:170. Tillich distinguishes three elements in life defined as the actualization of potential being, all of which reflect the centeredness of the person: self-integration or self-identity; self-alteration or self-creation; and self-transcendence and return to the self. ST, III:30–31.
12. Tillich, ST, I:170.

13. Paul Tillich, *A History of Christian Thought*, ed. by Carl E. Braaten (New York: Simon and Schuster, 1967) 457.
14. Tillich, ST, I:184.
15. Tillich, ST, II:32.
16. Tillich, ST, I:184–185.
17. Wolfhart Pannenberg, *Anthropology in Theological Perspective*, tr. by Matthew J. O'Connell (Louisville: Westminster/John Knox Press, 1985) 57; see: 498–500.
18. Pannenberg, *Anthropology*, 73.
19. Pannenberg, *Anthropology*, 527.
20. Pannenberg, *Anthropology*, 528.
21. Karl Barth, "The Gift of Freedom: Foundation of Evangelical Ethics" in *The Humanity of God* (Louisville: Westminster/John Knox, 1968) 75.
22. Langdon Gilkey, *Reaping the Whirlwind: A Christian Interpretation of History* (New York: Seabury, Crossroad, 1976) 235.
23. Polkinghorne, *The Faith of a Physicist*, 81. Arthur Peacocke says, "*God has a self-limited omnipotence and omniscience.*" *Theology for a Scientific Age* (Minneapolis: Fortress Press, enlarged ed. 1993) 121, italics in original.
24. Wolfhart Pannenberg, *Theology and the Philosophy of Science* (Louisville: Westminster/John Knox, 1976) 302.
25. Karl Barth, *Church Dogmatics*, 4 Volumes (Edinburgh: T. & T. Clark, 1936–1972) II/1:301. Hereinafter CD.
26. Barth, "Gift of Freedom," 77.
27. Barth, CD:II/1:313.
28. Barth, CD/1:309. Barth is a theist, and as such takes a strong stand against both pantheism, according to which God's being is exhaustively present in the natural world, and panentheism, according to which God's being is present in the world but the world does not exhaust God. "God enters into the closest relationship with the other, but He does not form such a synthesis with it." Ibid., 312.
29. Barth, "Gift of Freedom," 78. God's liberation gives us "redemption for responsibility." CD:IV/3:663.
30. Barth, CD:IV/3:660–661. When referring to God's action to redeem, Barth prefers the term "liberation" because it is more dynamic compared to "freedom" which is more static. Ibid., 663.
31. Russell, *Human Liberation in a Feminist Perspective*, 42.
32. Letty M. Russell, *The Future of Partnership* (Louisville: Westminster/John Knox Press, 1979) 102.
33. Elsewhere I spell out in more detail the structure of proleptic ethics. *GOD—The World's Future* (Minneapolis: Fortress Press, 1992) Chapter 12.

Appendix B
Playing God with David Heyd

1. David Heyd, *Genethics: Moral Issues in the Creation of People* (Berkeley and Los Angeles: University of California Press, 1992) 4.
2. Heyd, *Genethics*, xii.
3. Heyd, *Genethics*, 218.

Index